"Writing beautifully about our most unbeautiful cognitive apparatus, [Broks] describes its rumpled surfaces, its network of neurons, its deep, secret spaces with such care that the act itself powerfully illuminates the brain's creative capacities."

—Lauren Slater, *Elle*

"British neuropsychologist Broks, called 'the next Oliver Sacks' by the London press, actually proves himself to be more freewheeling and personal than his acclaimed precursor with this collection of philosophically inclined case studies."

—*Time Out New York*

"[Broks] bears comparison with such accomplished physician authors as Oliver Sacks and Atul Gawande. . . . [He] displays skill at combining disparate elements . . . and a variety of writing styles and ideas, into a surprisingly satisfying narrative."   —*American Scientist*

"An intriguing investigation of the link between the brain and the mind . . . In fragmented case studies, imaginary conversations, short stories and essays, [Broks] explores how the physical structure of the brain translates into what we know as the self. Readers who enjoy the case studies of neurologist Oliver sacks will be drawn to this book."

—S. M. Colowick, *Olympian*

"In an engaging series of vignettes about neurologically impaired people, Broks eloquently details how he and his colleagues analyze the relationship among personality, performance, and grain anatomy."   —*Science News*

"A tour-de-force intertwining of the clinical, the personal, the fictive, and the philosophical that . . . certainly keeps the pages turning. . . . Broks writes like an impressionist painter, splashing his canvas with vivid colors that capture a moment with emotional force and mystery."   —*Kirkus Reviews*

"Broks uses cases as jumping-off points for essays in search of personality, unique consciousness, the soul."

—Ray Olson, *Booklist*

"*Into the Silent Land* is a small, strange, beautiful gem—a brilliant lattice of arresting neurological tales, hard-nosed contemplation, and, unexpectedly, a certain wistfulness. Broks is as much poet as scientist, and in this indelible book, he leads us effortlessly into an unfathomable mystery—how that pale substance we call the brain could create something so ethereal and individual as a human mind."

—Atul Gawande, author of *Complications*

Into the Silent Land

Into the Silent Land
Travels in Neuropsychology

PAUL BROKS

Grove Press
New York

First published in Great Britain in 2003 by Atlantic Books, an imprint of Grove Atlantic Ltd.

Printed in the United States of America

The author and publisher wish to thank the following for permission to quote from copyrighted material: "In the Theatre" from *Collected Poems,* by Danny Abse, copyright © 1977. Reprinted by permission of the University of Pittsburg Press; the estate of Gilbert Ryle for *The concept of mind* (London: Hutchinson, 1949).

Earlier drafts of a number of these chapters have appeared in *Prospect* magazine.

"The Seahorse and the Almond" was published, in earlier form, in *Granta* magazine.

FIRST GROVE PRESS EDITION

Library of Congress Cataloging-in-Publication Data

Broks, Paul.
    Into the silent land : travels in neuropsychology / Paul Broks.
        p. cm.
    ISBN 0-8021-4128-5 (pbk.)
        1. Neuropsychology—Popular works. 2. Psychophysiology—Popular works. 3. Broks, Paul. I. Title.

QP360.B7956 2003
152—dc21   2003041854

Grove Press
841 Broadway
New York, NY 10003

04 05 06 07 08   10 9 8 7 6 5 4 3 2 1

For Sonja, Daniel and Jonathan

The brain is wider than the sky,
For, put them side by side,
The one the other will include
With ease, and you beside.

Emily Dickinson

CONTENTS

THREE   No Water, No Moon

ONE

Swallowing the Dark

Different Lives

'Why does raw meat give me a hard-on?'

This is Michael, chopping sirloin ready for the stir-fry. Typically, he is going to the trouble of preparing a good lunch: beef in hoi-sin sauce. He's bought some beer, too. We're drinking straight from the can. Amy, his girlfriend, sits at the kitchen table reading a magazine.

'Michael,' she says, without looking up.

Michael slides the diced beef into the wok where it sizzles in the hot oil.

'Easy, Amy. Only a twitch.' He winks at me, then drops what he is doing and strides out of the room. 'Have a listen to this,' he calls over his shoulder and soon the place is awash with cascades of sound – brittle arpeggios, tumbling fragments of melody. It is very loud.

Michael returns, fingertips to temples, head tilted back. 'Koto,' he says. 'Japanese. Astonishing.'

From this angle the dent in his head, about three inches up from the right eyebrow, is more noticeable.

Next day I'm over at Stuart's. We sit in his stuffy front room. An ornate black clock (his early-retirement present) clings to the wall like a huge fly. As I struggle with milky tea, Stuart locks me in his gaze. He is about to say something, but doesn't. It is a long pause. Eventually he speaks.

'I don't love you any more, do I, love?'

The words are intended for his wife, Helen, who sits beside him. 'No, love,' she replies. 'So you say.'

There is silence again, except for the tick of the insectoid clock. The dent in Stuart's head is above the left eyebrow.

Michael had climbed a tree to retrieve an entangled kite. He needn't have bothered because the wind gusted and the kite drifted down of its own accord, but he was high up by then. He was calling something to Amy, but she couldn't make it out. Her dreams recall how abruptly his voice was stifled by the creak and crack of a branch, and the wind-whipped silence of the free fall as his body cleared the boughs. Concealed within thick tufts of meadow grass was a spur of rock. Amy's dreams also record the crack of head hitting stone. That's what wakes her.

The fall fractured Michael's skull and released a flash flood of bleeding into the right frontal lobe. 'I thought his number was up,' the surgeon told me, and had said as much to Amy as she kept vigil over Michael's comatose body. 'No point beating about the bush,' said the doctor. But, after three days and nights, Michael came back to life – with a different number.

Stuart's twist of fate was a motorway pile-up. A bolt snapped and blasted like a bullet from the vehicle in front. It came through the windscreen, through his forehead and tore deep into the left frontal lobe.

Despite the immediate displacement of some brain matter, loss of consciousness was brief, as is sometimes the case with penetrating missile wounds. He told the paramedics he was fine and had better get home now, but they saw the brain stuff gelling his hair and put him in the ambulance. Soon the surgeons were working to extract the foreign body from the interior of Stuart's head, a process that also meant disposing of some adjacent brain tissue. Part of Stuart went with it.

By these means, Providence has created mirror-image lesions of the brain. As a neuropsychologist, my role is to compare the consequences. Stuart now has trouble getting started. Helen encourages him out of bed in the morning, points him in the direction of the bathroom, has his clothes ready, and gets him breakfast before going to work. She leaves him lists of things to do around the house, and magazines and puzzle books to fill the hours. But when she returns she often finds him where she left him, sitting in silence. She'll go over and hug him and he'll return the embrace, but it's perfunctory.

He doesn't love her any more. It's the plain truth and she accepts it. Stuart is not to blame. What he feels towards Helen is what he feels towards all other people, including himself: indifference. This absence of emotion frees him to tell the truth: 'Helen, I don't love you any more.'

Stuart can read people's moods and motivations, but lacks the emotional charge of empathy. I ask what he feels about the little girl who was abducted and murdered last year. He knows it was a dreadful thing to happen. They should hang the murderer or chop his balls off but, no, it doesn't make him feel anything very much. Then, he says, it's funny but he never used to believe in capital punishment.

Michael, on the other hand, has trouble stopping. Amy has to rein him in. He'll talk to strangers in the street, he'll tell them they're beautiful, or their children are, or their pets. He wants to touch. He wants to celebrate. Beggars bring a tear to his eye. He once gave a man his coat and a £10 note. People take advantage.

Michael's empathic response is hair triggered, but more complex social calculations befuddle him. When he first came home from the rehab centre his tastes were plain. Amy said he lived on fish fingers and Led Zeppelin. Michael said it was like going back in time. He'd always liked these things and now he didn't feel he should pretend otherwise. Fine, said Amy. But she would not tolerate the porn videos. Like Stuart, Michael no longer feels the need to dissimulate.

'How do you feel in yourself, Stuart?' I ask.

'All right.'

'Are you miserable?'

'No.'

'Are you happy?'

'I don't think so.' He turns to Helen. 'Am I happy?'

Helen looks at me. I look at Stuart. The question goes round in a circle.

Michael saw me off at the front door. He was close to tears. He pulled me to him and kissed me on the cheek. For an instant I thought he was going to say he loved me.

* * *

The bald head swivels. The voice honks like a klaxon across the senior common room: 'I never make mistakes.' There is a rustling of newspapers and clearing of throats.

Martin has superior intelligence – my tests confirm it, and he holds a master's degree in mechanical engineering – but he happens to be autistic and has a problem with volume control. Is that a reason to bar him from the SCR? No. We'll enjoy our coffee.

He's been doing one of his party pieces: calendar calculation. Martin can give you the day of the week for any date you care to mention, and he's spot on every time, seldom taking more than a couple of seconds. He's happy to oblige and seems disappointed when I soon run out of dates I can vouch for.

'How do you do it, Martin? You didn't even think about the last one.'

The target date was 18 March 1988 (my son's birthday). 'Friday' was the instant response.

'That was easy,' he says, 'I went to the dentist the day before.' He grins with satisfaction.

It's hard to tell his age. The face is lined but unweathered. He's wearing a silver puffa jacket, sta-pressed trousers at half-mast, and trainers. Forty-eight going on fourteen. That should be 'trainer' in the singular. It's on his right foot.

'I see you're wearing odd shoes,' I say.

'Yes,' he replies. 'It's Wednesday.' I wait for further explanation, but none is forthcoming.

When I first saw Martin, for clinical assessments, he turned up with his parents and they'd put him in a suit. His shoes were polished, and matched. He hardly said a word. Today, in his casual attire, he is voluble. Before long, inevitably, he drops into the groove of his special interests. There are several. One is the Beatles. He knows the recording and release dates of every record. Another is the railways. He has memorized the regional timetable, of course, but what really fascinates him is the movement of coal freight wagons. Then there is astronomy, which, currently, is his main preoccupation.

'Do you know how many stars there are in the universe?' he asks. 'Ten to the power of twenty-two.'

I make a little blowing sound and shake my head. He looks pleased.

'Actually,' I say, 'I read somewhere that if you think of each star as a grain of sand it would take all the beaches and deserts on the planet to match the number of stars in the universe.'

I thought this would impress him, but he ignores me. He becomes agitated, starts rocking back and forth on the edge of his seat. When he stops he says, 'I don't think so.'

I ask him if he thinks there is intelligent life out there among all those grains of sand. He looks puzzled and I realize he's taken the question literally, so I clarify. Again, the grin.

'Yes,' he says, 'there is.' The smile is sustained. It is evidently a consoling thought.

Beth joins us. She's one of our research assistants. It's time to go to the lab for the testing session. Martin's face lights up. He has taken a shine to Beth.

'And what have you been up to?' she asks him.

'I've been masturbating quite a lot,' he replies, as if through a tannoy. I press mouth against knuckles to block the laughter. It's no good. I snort and cough.

'Excuse me,' I say and cough again for good measure. It's unprofessional, I know, but he cracks me up. I'm only human. I'm not trying to make Martin look ridiculous. He *is* ridiculous. Look at him in his daft clothes, booming on about masturbation and coal freight wagons and the number of stars in the universe. It's undeniable. And I reckon it's a snub if you *don't* acknowledge his absurdity. If you are to engage with Martin you must, to some extent, enter his world.

'Martin,' I say. 'This is funny. Do you mind if I laugh?'

'No,' he says. 'Please laugh.'

But, given permission, I find the humour soon dissolves, and I'm left sitting red-faced with tears on my cheeks and everyone looking at me instead of him. I even find myself pondering Martin's confident assertion of the existence of extraterrestrial life. *We are alone in the universe or we are not*, I think. *Either way, how astonishing*. We grin at each other.

His head is abnormally large, as is the brain that fills it. My colleagues and I have taken measurements. We are profiling his cognitive strengths and limitations and setting these against detailed magnetic resonance observations of his brain. He is an enthusiastic research participant and has come to see himself as a neuro-engineering problem.

He has a theory. In his view autism is all about flow dynamics. Most of the time his thought processes are stuck in the left hemisphere of his brain. Consequently, his thinking is rigid, categorical, and analytic. If he could unblock the channel of the corpus callosum, which links the two sides, then the streams of the left and right brain would merge and he would be whole. Ordinary consciousness would flourish. This happens sometimes, he believes. For brief periods the world takes on a different appearance. He is more relaxed and it is less of an effort to connect with people. This is where masturbation comes in: orgasm detonates a dam-busting explosion in the right hemisphere.

As Beth sees Martin to the door, I catch a fragment of their conversation.

'But if your boyfriend leaves you . . .' he says.

'We'll see,' says Beth.

Martin's grin has an unworldly beauty.

* * *

It was her seventh birthday, Ellie's father is telling me, a clear morning in April. They had stopped to chat to a neighbour. Ellie was losing patience. She wanted to ride her new bicycle. He can see it now, blue and silver chrome, dazzling in the sunlight. And then, 'She was lying in the middle of the road, dead still. It was like the world had stopped, except for me. When I got close the rest caught up; the screech of tyres, the bicycle scraping across the road. Someone said, "Oh my good Lord!"'

His mind held a contradiction as he looked down on his daughter's body: *She's not badly injured* and, at the same time, *She's dead*. Neither was the case. Not the latter because, of course, she is here, a young woman now, squeezing his elbow; and not the former. Her arms were grazed, nothing serious, and her face was unblemished. But what her father could not see was the fractured parietal bone and the slow seepage of blood into the right hemisphere of Ellie's brain.

It would be a week before she opened her eyes. But she was not dead. And through the tunnel of intensive care – she in coma, he consciousness flayed – Ellie's father found the strength not to pray. His prayerless vigil was rewarded. Ellie recovered and, months later, returned to school. He dropped her off at the gate and says he blubbered so much on the drive to work he had to stop the car. Joy can be so profound it borders on grief.

Ellie never regained the full strength of her left arm and leg, and she tired easily, but it didn't stop her joining in with the other children. She struggled to concentrate and keep pace in some lessons, but that was to be expected. No one pushed her; she pushed herself. She found a talent for languages and is now preparing to go to university. So what's the problem?

'Parallel parking,' says Ellie, 'and overtaking.'

She has difficulty judging speeds and distances. She's twice failed the driving test. Is it anything to do with her brain injury and, if so, can I help?

I finish my assessments at the next appointment. Ellie has worked hard at tests of spatial awareness, motor co-ordination, concentration, and reaction time. The results show problems consistent with her brain injury. She senses this and, with a kind of desperation, offers to take me for a drive. I accept.

'Do you want me to come?' asks her father.

'No,' I tell him, 'go and have a cup of tea.'

At first Ellie seems unsure where the car is parked. It's an old Citroën, the colour of tomato soup.

'Where shall I go?' she says.

'Anywhere. Just drive around. Go left here, then next right.'

And so we go, me giving directions. I have to admit she's pretty good. Ten minutes into the drive nothing untoward has happened and I'm beginning to question the value of my tests. There's no doubt she had problems, but here we are in the real world and she's doing fine.

Ellie has steered the car into the middle of the road ready to turn across the oncoming traffic back into the hospital car park. The indicator clicks as we wait. It's a comforting sound. *Tick, tick, tick.* Almost hypnotic. There's a steady flow of traffic, so Ellie waits. *Tick, tick, tick.* Then a gap; nothing for fifty yards, space enough to get across. But we don't move. *Tick, tick, tick.* Another line of traffic draws towards us, headed by a white removals van with YOUR MOVE! splashed across the front. *Tick, tick, tick.* YOUR MOVE!

The image of the van now filling my retina and flashing into my brain takes the quick-and-dirty route via the thalamus and straight to the security monitors of the amygdala, deep in the temporal lobe. *Action stations!* No need to trouble the higher cortical centres just yet, because something has impelled Ellie to turn across the traffic and we are going to hit the van. Conscious deliberation would be a hindrance. This is basic survival. My arms fly up and my head jerks sideways. The amygdala screams instructions to the brain stem, signalling the release of chemicals into the bloodstream and, through a clatter of synaptic activity, galvanizing the autonomic nervous system. This is *red alert*!

Then I become aware of the pig squeal of tyres – the van's, not ours. My cortex is coming back on-line; reflective consciousness restores itself. We roll serenely on and I glance back to see the van pulling away. Ellie remains unperturbed.

We are back in my office. 'It was a close call,' I say.

'Oh?'

'I thought that van was going to hit us.'

'What van?'

I tell her that we could arrange for a more advanced assessment of her driving skills, and that she is obliged to inform the driving licence authority of her condition.

'I already have,' she says.

But I can't encourage her to drive. I see a damaged brain encased in a tonne of metal cruising down the motorway, through rush-hour traffic, through residential areas where children are riding their birthday bicycles. The damage is beyond repair.

'I came to you for help,' says Ellie. Her father gives me an empty 'Thank you' as they leave.

A few months later I get a call from Ellie. She has taken her driving test for a third time and passed. 'I thought you'd like to know,' she says.

I picture her father standing beside her. What's that on his face? Absolution?

* * *

Mrs O'Grady is showing me photographs. There are three albums opened out on the coffee table. There she is at Katie's wedding; small, nervous, and neat in a pale green suit. Two months on, there she is at Stephanie's. Beige this time.

'I feel guilty,' she confides. 'I still haven't told Steph. Do you think I should?'

'Yes,' I say. 'She'll understand.'

I decline a second cup of coffee and gather my stuff to leave, but I'm not going yet because Mrs O'Grady has grabbed my arm. She leads me to a corner of the room and stands back with an air of curiosity. She stares, steps forward, stands back again. She can't make me out. The facial musculature shapes apprehension, building to dread. Then she goes blank.

She walks to the other side of the room, smacking her lips and tugging her collar. I follow her to the kitchen where she stands by the stove picking her nose. Then she fills the kettle, but doesn't switch it on. She fetches mugs from the cupboard and places them on a tray. From time to time she seems to be aware that there is someone else in the room. She looks at me, but I am too much to fathom. I feel semi-transparent. I speak, but there is no response. Am I really here?

She fills the mugs with cold water from the kettle and carries the tray into the living room. We sit in silence. I'm thankful this hasn't developed into a thrashing, foaming, full-blown fit. After a while she reaches for the third album.

'This one's the holidays,' she says. 'Tenerife.' But she knows something is wrong when she sees plain water in the mugs.

Mrs O'Grady takes brief excursions from consciousness. These are known as automatisms, a feature of her epilepsy. The conscious mind switches off, but the bodily apparatus carries on in a more or less purposeful fashion: feeding the cat, walking round the supermarket, boarding a bus. Had she reached for the bread knife and plunged it through my heart I doubt she would be convicted of murder. The law makes provision for automatisms.

Watching Mrs O'Grady's unoccupied body scuttling about I thought of her as a zombie. Students of consciousness are fond of zombies. Not the Haitian living dead or shambling ghouls of the *Twilight Zone*, but far stranger inhabitants of the world of philosophical conjecture. These creatures look and act like ordinary people; they walk, talk, sing, laugh, and weep, have love affairs, raise families, get drunk, argue about politics. They are, in fact, like us in every way but one: they lack conscious awareness. Their brains regulate internal states of the body and control outward behaviour, but that's all. While the rest of us move about in a bright pod of consciousness, zombies just move about. Their philosophical purpose is to crystallize the mind-body problem. Is it logically possible to subtract mental life from the working brain, in which case there would be scope for zombies (dualism)? Or are brain activity and consciousness one and the same thing (materialism)? No doubt Mrs O'Grady would have something to say on the matter.

The trouble is, not all of her excursions are so brief; hence Mrs O'Grady's guilt over Stephanie's wedding. Her memory holds no trace of the occasion. Physically she was there. You can see her in the photos. But she was not there mentally, at least not in full. It was too protracted an episode to fit the conventional scheme of an epileptic automatism. More likely her brain had settled into a stable pattern of dysfunction with low-level epileptic discharges jamming the transmission of sensory information into memory. Her awareness would have been a fragile membrane of impressions floating between 'now' and 'then', but never quite connecting.

There are other circumstances in which human beings appear

to act purposefully without the benefit of self-awareness. Sleep-walking is a good example. I was in the Combined Cadet Force in my teens. One night, at camp, I somnambulated through the barracks and mistook the NCOs' quarters for the lavatory. I shuffled in and urinated over one of the officers as he slept. Unfortunately, the following morning I was fully conscious.

How convenient it would be sometimes to turn off consciousness and carry on with ordinary behaviour. Imagine flicking a switch on difficult days and flipping into oblivion, knowing that your body will continue going about its normal business. No one would notice. A pre-programmed wake-up call would return you to sentience in time for a film or the football. Controlled automatism might be preferable to periods of physical or emotional discomfort, or sheer boredom. If everyone had a consciousness switch then the world, most of the time, would be teeming with zombies. Perhaps it already is.

What troubles Mrs O'Grady is that she remembers one wedding and not the other: Katie's, but not Steph's. It seems unfair. In truth, she says, it's not so much that she can't remember as the feeling that she wasn't actually there. Like she didn't bother to turn up. I'm not going to debate it with her and, for her own peace of mind, I think she should talk it through with her daughters. But if they couldn't tell, what difference does it make?

Later, lying in bed, I confess to my wife that I am a zombie. We had a malfunction with the transcranial magnetic stimulator. It zapped my awareness module. I thought she should know, but best not break it to the kids just yet. I say I hope it won't change the way she feels about me. She is already asleep.

The Space behind the Face

The illusion is irresistible. Behind every face there is a self. We see the signal of consciousness in a gleaming eye and imagine some ethereal space beneath the vault of the skull, lit by shifting patterns of feeling and thought, charged with intention. An *essence*. But what do we find in that space behind the face, when we look?

The brute fact is there is nothing but material substance: flesh and blood and bone and brain. I know, I've seen. You look down into an open head, watching the brain pulsate, watching the surgeon tug and probe, and you understand with absolute conviction that there is nothing more to it. There's no one there. It's a kind of liberation.

The illusion is irresistible, but not indissoluble. It is more than twenty years since I began my clinical training at a rehabilitation hospital for people with neurological disorders. I was a student of clinical psychology, but was drawn mostly to neurology. For as long as I could remember I had been interested in the workings of the brain and, one way or another, as clinician or scientist, I expected to make a career in neuropsychology, the science of brain and mind. Neuro-rehab was a good place to start.

One of the patients was a seventeen-year-old boy who had stepped into an empty lift shaft through which he fell three floors, almost to his death. The surgeons had done their best to piece him together again, but now the dome of his shaven head was asymmetrical: convex on the right, concave on the left, with a deep oval depression like the shell of a hard-boiled egg cracked with a spoon.

His face worked relentlessly, writhing with anger and dread. Mostly anger. He would growl and grunt, and sometimes howl, but, apart from occasional volleys of obscenity, he was incapable of speech. This is not uncommon. People without ordinary speech due to brain injury sometimes have the capacity to summon up the vilest gobs of abuse. I didn't know that at the time. It came as a shock. Sometimes they can also sing, but this boy never sang. He sat contorted in his wheelchair, head turned sideways and back at an uncomfortable angle, limbs buckled with spasticity, a stream of saliva dribbling from the corner of his mouth.

The priapism was a final, humiliating twist. Due to a quirk of the damage to his nervous system he was continually troubled by a painful erection. The young women tending him – nurses, physios, and occupational therapists – pretended not to notice.

I felt pity for him, but also revulsion. As a raw trainee not yet acclimatized, I found him grotesque. What disturbed me most was the flickering screen of his face: bleak images of a soul in torment. Or so I imagined. Then I began to consider what might remain of a 'soul' or a 'self'. I began to doubt there was anything at all going on behind that face. *He should be allowed to die*, I thought, and not just for his own sake. How did he look to his mother? Could she even bear to look?

The chaos of his face drained my sympathy. It broke the rules. A face should allow public access to the private self. It's an ancient convention of the human race. There is a universal system of signals. But this young man's facial displays worked like a subterfuge, denying knowledge of what lay behind. Perhaps nothing lay behind.

Then, one day, I happened to be around when the boy's mother came to visit. I watched as she cradled his broken head in her arms. For the time that she was with him, but not much longer, an extraordinary transformation came over his face. It became still. The rage subsided. He seemed to regain his humanity. Here were two selves, not just a mother and a broken shell of a son. The whole was greater than the sum of its parts.

Maybe it was a failure of imagination that led me to sense a seeping away of the boy's self once his mother had gone, but the capacity I discovered in myself to see a fellow human being as less than a person was an appalling revelation. In such circumstances how are we to distinguish failure of empathy from valid observation? Perhaps they amount to the same thing.

Now my own son has turned seventeen, the same age as the eggshell boy. He was in the void of pre-birth when our friend made his lonely descent through the lift shaft. To disturb someone from a state of non-existence is a terrible responsibility. Look at what can happen.

I have a memory of being with my son when he was four

years old. It is deep winter. We have to go out, so we leave the warmth of our house for the freezing night air. There are few lights in the village and the sky is full of stars. We are hardly beyond the front door when he starts coughing.

'Are you all right?'

'It's okay,' he says, 'I think I just swallowed some dark.'

He has the notion that darkness is a substance. It will make you choke if you swallow too much in one go. I could have put him straight with some prosaic account of the coughing reflex being triggered by the shock of the cold air rather than a mouthful of darkness, but I didn't. I stashed away the treasured image and left him with the version of reality fashioned by his infant brain.

Reality is under constant review. Twenty-three centuries ago, Aristotle believed that the heart must be the source of mental life because of its dynamic action and its warmth. The function of the brain, he thought, was to cool the blood. He built his cosmology on the belief that the Earth stands motionless at the centre of the universe, a fixed point about which the sun and the moon and the stars revolve. Aristotle was wrong on every count, but his erroneous beliefs – the product of intuition and illusion – served him well enough. And though we now know immeasurably more than Aristotle about the workings of the human body and the structure of the cosmos, we should not delude ourselves by thinking that we have arrived at some privileged end-point of intellectual evolution.

We still live by intuitions and illusions, especially when our thoughts turn inwards. The bright, intangible qualities of subjective experience have yet to be reconciled with the dark substance of the brain, but that space behind the face is still lit by the mind's eye. Irresistibly, we still see the vision of minds in the light of other people's eyes. Cosmologies come and go, but if this illusion begins to fade then so does the observer.

* * *

I've been trying to think of the eggshell boy's name. I could have given him a name. All the others have pseudonyms. It wasn't deliberate. I didn't choose to deny him a name. But when the story was finished, I saw he didn't have one. 'A name would humanize him,' someone said. 'Call him John or Steven or Richard . . .'

On reflection, I thought it might do just the opposite.

The Seahorse and the Almond

Whisky on top of wine was a mistake. This morning it has left me feeling fractionally too embodied; too aware of the weight and movement of my head, the bulk of my tongue.

I woke late, breaking from a thick crust of sleep not much before eight. Half an hour later, I'm walking to work, without hurry, but keeping pace with the traffic. It's a couple of miles. It will do me good. Down past the parade of shops, past the odd juxtaposition of casino and funeral parlour, past the terraced houses at the fringe of the park, and on up the other side of the urban valley towards the drab monolith on the brow of the hill. The District General Hospital is visible from most of the city. Today it is framed by a sky the colour of cement.

Naomi is deep inside. It is her nineteenth birthday. She is on a bed being pushed by a porter along shiny floors, into lifts and out across more shiny floors. She is tired, having been awake since the break of day, well before the neurophysiology technicians came to glue electrodes to her scalp. They left her with a Medusa's head of angry serpents.

Arriving at nine, I go straight to the angiography suite where preparations are in hand for Naomi's ordeal. The central chamber is small, about the size of a suburban living room. It is brightly lit and crammed with X-ray equipment, monitors, and control panels. The centrepiece is the narrow bed upon which the patient, when she arrives, will be laid. The way it tapers at one end reminds me of an ironing board. In the corner a quiet man from Medical Illustrations is setting up his video camera ready for the show. EEG technicians in white and radiographers in blue filter in and look busy.

We are going to interfere with the workings of Naomi's brain, anaesthetizing each hemisphere in turn with injections of Amytal, a fast-acting sedative. Our aim is to isolate and interrogate one side of her head and then the other. Strictly, 'anaesthetize' is incorrect since the brain has no sensory receptors. It is always in a state of anaesthesia.

The radiologist appears. 'Do we have a patient?' We do. Naomi is sitting up in her mobile bed, which has been parked just down the corridor, nowhere in particular. It has arrived as if by time-lapse photography moving from one indeterminate station to the next, and now here she is. She looks lonely, so I go and chat with her for a while. I wish her happy birthday.

I like Naomi. I've got to know her well these past few months, watching her progress through an obstacle course of investigations (EEG, MRI, video telemetry, neuropsychology) that will lead, she hopes, to the surgeon's list, to the operating theatre and to the carving away of a small streak of scarred brain tissue – the source of her epilepsy.

Her faith in the doctors and surgeons is absolute. The fits will cease. All will be well. And when she is seizure-free she will go to university. Perhaps she will take a gap-year and travel to Australia. In time, she will apply for a driver's licence. And so on. She is incorrigibly optimistic. It may be a feature of her brain pathology.

Her boyfriend, whose name I have forgotten, is less sanguine. Unlike Naomi, he can see the possibility of failure. He understands that the operation might not work. 'You're *such* a pessimist,' Naomi told him when I saw them both in the clinic. I'd be the same. *Be troubled, Naomi. A little. The surgeon, if he gets his hands on you, is going to open your head and take a piece of you away. Too much faith and expectation can be counter-productive.* I think these things, but this is not the time to voice my concerns. It is a time for unconditional reassurance – not my strongest suit, but a necessary part of the repertoire, and well practised.

Meanwhile, the radiologist is sifting through his tray of paraphernalia and realizes something is missing. 'Do we have any Amytal?' No, not yet. Our batch of the stuff looked suspiciously cloudy, possibly contaminated. No problem. A call to Pharmacy and we are assured that a supply of the drug is already on its way from the Radcliffe Infirmary. Why it has to come all the way from Oxford I've no idea. I don't enquire.

This morning's procedure – a Wada test – is the final hurdle. If Naomi passes the test she advances to the surgeon's list. She is prepared. Yesterday she rehearsed the protocol with one of my colleagues from the Neuropsychology Unit. She lay on her back, raised both arms to the vertical, counted up to twenty, imagined (at around ten) that her left arm had become limp and let it drop to her side. This is what happens when the drug hits the right side of the brain. Then they went through the motions of testing. Naomi performed some simple actions ('Touch your nose . . . close your eyes . . . blow . . .'); recited the days of the week and counted backwards from ten. She described a picture ('A man up a ladder, a boy with a ball, a girl, a kite, a dog and a cat, a pond, ducks . . .'); named objects; read sentences; did mental arithmetic. The instructions and questions were rapid-fire. Amytal is fast acting, but its effects are short-lived. In the test proper the injected half-brain will sleep for just two or three minutes – while we conduct our business with its wakeful twin.

My colleague arrives carrying a clipboard, a stopwatch, and two black ring-binders. On her way into the angio suite she exchanges smiles and words with Naomi, whose bed has now moved closer to the main door. There's a smile for me too. She already knows about the delay with the Amytal. Time for a coffee. We sit next to the machine that spills out the X-ray negatives and flip through Naomi's case notes. Her history is unremarkable. It all started with a fever when she was small. She'd been off colour for a couple of days, then seemed to pick up. Her mother wasn't sure, but in the end dropped her off at nursery school on the way to work.

Midway through the morning Naomi fell asleep in the sand-pit, or so the teachers thought. When she couldn't be roused they called an ambulance. She started shaking before she fell asleep, the other children said. The doctor thought it was probably a febrile convulsion: not to worry, a lot of kids are prone to seizures if their temperature climbs too high. They mostly grow out of it. And so, it seemed, she did. But the fits returned on the first tides of menstruation.

They were shadowy figures with a pungent smell of electricity, a sensed presence, but no one there. Odd to identify the smell of a seizure with electricity, which is odourless, but apt for an electrical storm in the brain.

The ethereal visitors are part of the epileptic aura, a state of altered awareness that serves to forewarn of an approaching seizure. It also has another, more visceral, feature. Naomi says it feels like a sparrow fluttering its wings in the pit of her stomach. The bird ascends to her throat, becomes trapped, and struggles to escape. Up to this point, under the gathering gloom of the brainstorm, in the company of the empty shadows and the sparrow, she is fully conscious and can articulate her experiences. Then the storm breaks and she is swept beyond reflection. Her eyes become glazed and empty. She tugs at her clothes, smacks her lips, and keeps wiping her nose with the back of her hand. I've seen her in this state. She has gone with the wraiths. They have left an automaton, acting out a purposeless, robotic routine.

After the tone poem of the aura – the unformed images, the unnameable scents – and after the rhythmic automatisms, there sometimes follows a third, catastrophic, movement. About one in five of her attacks turns into a generalized seizure, what would once have been called a grand mal. First, her muscles contract and she falls to the ground, sometimes spurting blood as her jaw clamps shut and her teeth sink into her tongue. She stops breathing and, unconscious, she urinates. Then come the convulsions – limbs jerking mechanically for several minutes – followed by release into a deep sleep.

Despite inventive cocktails of anti-epileptic medication, with dosages almost to toxic levels, the frequency of Naomi's seizures has steadily increased. Now she gets them almost every day. She is desperate for a cure and willing to take risks.

The planned operation has an ungainly name: amygdalo-hippocampectomy, so called because it involves removal of the amygdala (from the Greek for 'almond') and part of the adjacent structure, the hippocampus ('seahorse'). Each half of the brain contains an almond and a seahorse. The purpose of the Wada test is to clear a way for the operation. We know it is the right side of Naomi's brain that bears the scar tissue and drives the seizures because we've seen the brain scans and we've logged the clinical signs. But we are also making an assumption, possibly unwarranted, that her left hemisphere, which looks normal, is functioning normally. Our test will help determine whether this is true. (It is 'Wada', by the way, not 'WADA' as I've just been reading in the case notes; a common error. The procedure is named after Juhn Wada, the Japanese-Canadian neurologist who first proposed its use. It must be disappointing to be elevated to the status of an eponym only to be mistaken for an acronym.) We need to be as sure as possible that there is no 'silent lesion' on that healthy side; in other words, a malfunction that hasn't shown up on the brain scans. Appearances can be deceptive. Brain tissue can look clean and plump, but without putting it to the test one can't be sure of its integrity.

One of the targets for surgery, the hippocampus, is a vital component of the brain's memory circuitry, essential for laying down new traces. We need to know, above all, whether the left hemisphere of Naomi's brain is up to the task of sustaining basic memory functions. To the extent that each of us is the sum of our memories, the hippocampus is the instrument by means of which we assemble ourselves. Everything accessible to conscious recall has been registered and recorded through its channels.

What were you doing ten minutes ago? Who was the last person you spoke to? What did you have for breakfast? What did you do yesterday, last weekend? When was the last time you wept, and why? Conjure an image of your first school, the face of your teacher, your best friend. Remember your first kiss. And then, stretching to the mental horizon, rising through the squalls and sunshine of personal experience, picture the towering stacks of information in the public domain; the raw materials of culture. What does the word 'entropy' mean (or 'the' or 'word' or 'mean')? How do you use a telephone? Who is the president of the United States? At what temperature does water freeze? Who wrote *King Lear*? What is the function of the liver? All this information, personal and public, finds its way into memory by way of the hippocampus.

As an aid to recall, medieval scholastics developed elaborate, architectural systems of mental imagery – 'Theatres of Memory' or 'Memory Palaces' – through which they would take imaginary strolls, depositing or retrieving nuggets of information at strategic locations. I like the idea that the Keeper of the Gates is as fragile a creature as the seahorse. It doesn't take much – a stroke of the surgeon's knife – to finish it off and close the entrance for good. The flow of information stops.

If, as planned, the surgeon were to remove the right hippocampus, but it turned out that Naomi had no spare capacity in the left, then the operation would, in a sense, cause Naomi herself to stop. She would form no new memories of events or facts beyond her present age of nineteen. It would not prevent her from growing old, but her ageing body would forever house the mind of a nineteen-year-old girl.

Such things happened in the early days of epilepsy surgery. A handful of people, most famously a young mechanic known as patient 'HM', ended up with dense and irreversible amnesia, unable to retain new information for more than a few minutes at a time, and so unable to establish memories. You could visit HM every day for a year and each time he would greet you as a stranger. Leave the room for ten minutes and on your return he would have no idea who you were.

Since then surgeons have restricted their interventions to just one side of the brain, but even so there have been similar disasters where it was not established prior to surgery that the other side was in good working order. That's the reason we're here today, I remind myself, going through these arcane rituals. We want Naomi to continue in mind as well as body.

If the hippocampus is the gateway to memory, one can picture the amygdala as housing the levers of emotion. It links the information-processing activities of the higher, cortical areas of the brain – the machineries of language, perception, and rational thought – to deeper, older structures concerned with the regulation of emotion and motivation. In short, it tells us how to feel about what we are thinking and perceiving, and how to act on those feelings. Patients with damage to the amygdala on both sides of the brain inhabit a world devoid of emotional contour and colour. Diminished insight into their own feelings and behaviour is mirrored by a distorted perception of the emotional lives of others.

So the stakes are high for Naomi: memory and emotion. We need to get this right.

The Amytal arrives, delivered by motorcycle courier. He hands over a Jiffy bag that a nurse opens to find two phials containing a plain liquid, the stuff that will shortly work its spell on Naomi's brain. Our patient is now stretched out on the special bed, at the centre of things, waiting. Her head rests on a small square cushion. She is covered to the neck with a green surgical sheet except for an exposed patch around her groin, where, having administered a local anaesthetic and made a small incision, the radiologist is working to gain access to the femoral artery. Naomi's face at the top of the sheet and this framed expanse of pale flesh and pubic hair (and now blood from the cut) seem quite unrelated. Many people are surprised to learn that the most feasible route to the brain for these purposes is by way of the groin.

The catheter, a length of ultra-fine plastic tubing, is inserted and pushed, inch by inch, along the femoral artery, up through the abdomen and into the chest. Its journey is visible, magnified grainily in spectral shades of grey, on the X-ray monitors. I watch as it finds its way to Naomi's heart and from there to the junction with the internal carotid. She, too, is watching. She can see her insides on the monitors suspended overhead which, with exquisite integration of hand and eye, the radiologist uses to find his way from groin to gut to heart to brain. Next, a radio-opaque dye is pumped through the newly installed plastic piping to flood the blood vessels of the brain.

The radiologist takes a few X-ray snaps to confirm that we have reached our intended destination on the cerebrovascular map. Stationed at Naomi's midriff, he offers an occasional word of reassurance, and glances now and then in her direction. He means well, but the exchanges between them are perfunctory. She, for her part, is being a good patient. Her body is passive, receptive. Her face shows barely a trace of emotion, but when the nurse brushes a strand of hair from her forehead, Naomi's eyes moisten.

We have here Naomi the body, Naomi the mind, and Naomi the person. These, at least, are the differences of emphasis across the professional divisions of labour. The radiologist works in the realm of the flesh. He knows the intricacies of the vascular system and is on good terms with the ghosts of his X-ray machine. I, the neuropsychologist, will shortly signal a pharmacological invasion and deconstruction of Naomi's mind. The nurse, for now, is with Naomi the person.

We are all set to start. Naomi has her arms raised and begins to count. I look to my colleague standing opposite with her black folders, stopwatch, and clipboard, ready to assist with test materials and record responses. The neurophysiologists are a few feet back monitoring every squiggle of brainwave activity siphoned through the long bridal veil of multi-coloured leads attached to Naomi's head.

After a nod from me, the radiologist starts to inject the Amytal. It takes effect in a few seconds and I grab Naomi's arm as it swoons, guiding it to rest at her side. At that moment of collapse, the catching of the lifeless arm, something collapses inside me too and I catch myself thinking, *What am I doing here?*

I'd rather be somewhere else, well away from this unwholesome mind meddling. But here lies Naomi. There is work to do and, after all, better to be doing this to a relative stranger than to someone you love. That would be unbearable. She looks remarkably calm and ordinary given that her right cerebral hemisphere, half her brain, is now temporarily defunct. How *ordinary* she looks.

I am clear about the purpose of the test, but curious to know what is happening to 'Naomi the person', a question entirely peripheral to the immediate medical concerns. Our procedure is pharmacological, not surgical; the altered state is transient. But while the drug works its influence we have, effectively, amputated one side of her brain. I wonder if, with half of the brain closed down, we are engaging with just one half of the person.

Psychologists used to be obsessed by the duality of the brain. 'Functional asymmetry' was a hot topic when I was an undergraduate in the 1970s. The belief was that the two halves of the brain perform distinct, though complementary, functions: left hemisphere for language, right for spatial awareness; left for rhythm, right for melody; rationality / intuition; analysis / synthesis, and so on.

At the centre of attention at that time, scientifically and imaginatively, were the so-called 'split-brain' studies. Split-brain surgery was a radical method of treating people who suffered from severe and intractable epilepsy – patients tormented by frequent, debilitating seizures that could not be controlled by any other form of treatment.

Epileptic seizures are caused by abnormal bursts of electrical activity in the brain. The rationale for split-brain surgery –

commissurotomy – was that cutting the corpus callosum, the main channel of communication between the two hemispheres, would confine the abnormal electrical activity to one side of the brain and so prevent the development of major seizures.

I was not much concerned with the clinical aspects of this operation. I had no particular interest in epilepsy. What intrigued me was that the split-brain patients were *thought experiments made flesh*. They fell into the category of philosophical conundrum that also includes the 'brain in the vat', the 'brain transplant', and science fiction fantasies about teleportation and mind duplication.

Thought experiments are 'Imagine if . . .' scenarios designed to challenge our ordinary intuitions. In the seventeenth century John Locke explored the concept of personal identity by imagining an exchange of brains between a prince and a cobbler. It is psychological continuity that counts, he concluded. The prince 'goes with' his brain and now finds 'himself' in the body of the cobbler (and vice versa). More recent variations on the theme, some inspired directly by the split-brain cases, are less straightforward.

What if someone's cerebral hemispheres are divided and transferred separately (memories, character traits and all) to the heads of two different people? What if you swap a hemisphere with your best friend, or your worst enemy? Which of you is which? There would be some continuity in these cases, but not unity. Conventional notions of personal identity would be seriously challenged.

But the split-brain cases were not flights of philosophical fancy. They were real people with a real and irreversible surgical division of the brain. Inevitably they provoked as much philosophical interest as scientific. Like many others, my own imagination was captured by the suggestion that, in dividing the brain, the surgeon's knife was also dividing consciousness and therefore dividing the person. The very idea of bisecting the living, conscious brain clean down the middle was bizarre and absurd. It had a touch of the macabre, a whiff of the chamber of horrors. There are many weird creatures in the menagerie of neurological disorder, but the split-brain patients were of the purest strangeness. I was drawn in.

'Strange cases', closely observed, have an important place in the neurological literature. Alexander Luria, a major figure in the history of neuropsychology, was a master of case description and a persuasive advocate of 'romantic science'.

'When done properly,' he said, 'observation accomplishes the classical aim of explaining facts, while not losing sight of the romantic aim of preserving the manifold richness of the subject.'

I don't hesitate to recommend the popular writings of Luria, Oliver Sacks, and others to students as a way of introducing them to the field, but I recognize that part of the appeal, part of that 'manifold richness of the subject', has little to do with science or philosophy. It has more to do with the intrinsic fascination of the aberrant and the bizarre. *Morbid fascination* would not be too wide of the mark.

In this light, neurological case histories have a certain Gothic appeal. Replace the dark forests, the craggy mountains, the ruined abbeys, and the elemental storms of the traditional Gothic tale with a desolate urban landscape. Let a dilapidated modern hospital stand for the crumbling medieval castle with its labyrinthine passages, gloomy dungeons, and torture chambers. The white-coated, mad scientist in his cobwebbed laboratory, amid van de Graaf generators, lightning conductors, and the paraphernalia of alchemy, becomes the green-gowned surgeon in the sterile gleam of the operating theatre, knife in hand, ready to rework the brain's slimy fabric. At the centre of it all is the monster, waiting for the life force from the heavens to jolt its dead limbs, and the patient, brain exposed to the air, waiting for the blade.

And here I am now, in the shadow of Dr Frankenstein, having isolated one half of Naomi's brain, about to engage in a dialogue with . . . what? A person? A half-person? Half a brain?

It doesn't feel as if I'm dealing with some fragment of a mutilated self. Naomi's spirits seem to lift. She answers my questions obligingly and follows instructions with hardly a moment's hesitation. Those three short minutes fly by. 'You've done very well,' I tell her.

The drug wears off. The left arm has returned to life and the EEG trace is back to normal. Her eyes are closed and Naomi looks as if she is asleep. We know she isn't from the rhythms of the EEG – her brain is idling in a comfortable alpha rhythm, indicating relaxed wakefulness. It is time for the next stage of the procedure, to see whether she remembers anything from the drug phase. For Naomi, this element of the ritual is crucial. Failure here would outweigh success at any other stage. If she is to proceed to surgery she must pass my memory tests.

She's not doing so well. *Come on, Naomi, come on*, I'm thinking. 'The picture, Naomi, what can you remember?'

Perplexity, then a burst of information: 'A man on a ladder, a dog chasing a cat, a pond with some ducks on it, a girl with a kite, a boy with a ball.' All, unfortunately, from the picture she was shown at yesterday's rehearsal.

Formal testing completed, I ask Naomi how she found the experience.

'No problem,' she says. 'It was a breeze.'

The brain's language circuits are usually located on the left side, so disturbance or complete loss of speech is the typical response to injection of the left hemisphere. In other ways the effects are less predictable. Some patients appear confused and disoriented, some become agitated, some disinhibited. Others, like Naomi, just look desolate.

Her head is still, but her eyes flash left and right. She will not respond to my simple commands: 'Touch your nose, Naomi, touch your nose.' Nothing. When we get to Days of the Week she tries very hard, but all we get is 'Fa-fa-fa-fa-fa . . .' On Counting Back from Ten she approximates the number words, but gets locked in a perseverative loop: 'Tem, nipe, ape, ape, ape, ape . . .' She looks at the picture and has an urge to point at things: 'Da, da.'

For a while she seems to be warming to the task, appears engaged, but her concentration suddenly fades. At one point she looks me in the eye and chuckles wickedly, then another wave of emotion sends her in a different direction. Her eyes dart left and right again. She looks terrified; she looks *feral*.

'Fine, Naomi, just fine,' I say as we complete the routine. 'Relax, we're nearly there now.' We retire to a side room leaving Naomi to rest and recover from the drug.

The speech disturbance confirms for us that Naomi's language control centres are located primarily in the left hemisphere. This is important for the surgeon to know, giving him greater licence for excursions into the right temporal cortex if necessary, with minimal risk of disrupting language functions. And when it comes to memory testing there are no surprises. Her failure to recall or recognize most of the test items confirms that we were placing unreasonable demands on the damaged right hippocampus: taunting the crippled seahorse.

The one exception is her accurate recall of the mental arithmetic task. Under the drug she had stared at the sum printed on the test card and said, 'Sebber, seffen, fife, fife, five.' Now, correctly, she recalls, 'Four plus five equals nine.' I've seen this before. Somehow, I think, numerical information must gain back-door access to the left hemisphere in a way that verbal information cannot.

Recall of the experience of left hemisphere suppression is also less predictable than for shutting down the right hemisphere. Some patients have no recollection of events at all, at least nothing they can put into words. Others have at least partial insight into the frustrations of their temporary loss of speech. Some, like Naomi, just tell tales. 'Oh, it was okay,' she says. Well, perhaps she did have some slight problems with her speech at first, but after that she was fine. Maybe it was a bit different to the first time around, but not much, not really. Quite enjoyable. This is the left hemisphere confabulating. It does this for all of us, every waking moment. It edits our conscious experiences, makes them comprehensible and palatable. It is the brain's spin-doctor.

* * *

Two things disturb me during the night. One is the hoarse, *sotto voce* bark of an urban fox, receding in triplets down the street. The other is a fragment of dream, sharp enough to wake me. I stagger, giddy from being spun in a large machine they called an Accellotron. It has made me invisible, temporarily. I see my daughter sitting in the garden and approach her. I speak. She looks at me, but her eyes continue searching. I have, truly, become invisible. There's no way I can reassure her. She is terrified when I touch her hand. I'm terrified.

Next day, first thing, I'm sitting in my office watching the video of Naomi's Wada test. It's easy to miss important details in the patient's responses so we always check the video. There was something I failed to catch. The left hemisphere is suppressed and the origin of Naomi's fragmented, mumbling speech is uncertain. It could be the left hemisphere running on empty or perhaps it is coming from the other side of the brain. Either way it's hard to make out what she's saying. For a moment, her confusion seems to subside and there is a look of accusation in her eyes.

'Watafam,' she says, 'dooneer.'

I listen closely a second and third time and realize it's a question: 'What the fuck am I doing here?'

The Sword of the Sun

I had never been much aware of my father's glass eye, just as I had never really noticed his foreign accent. We were swimming some distance from the shore. Fourteen years old, I was way ahead. He called and I turned to find him treading water, right hand covering the empty cave of his eye-socket, good eye exploring the glimmering depths. The fugitive eye stared up at us from the seabed. I plunged like a pearl diver, following its gaze all the way down, and snatched it up with a handful of sand.

That evening, skimming stones into the sunset, I returned in imagination to the ocean floor. It was a cold and lonely place. Then it occurred to me that, deprived of an observing eye, the ocean was nothing. Such power! I closed my eyes and it was gone.

Years later I read Italo Calvino's *Mr Palomar*. It stirred memories. The eponymous hero goes for an evening swim. As the sun goes down it sends a dazzling band of light across the sea. Looking back to the shore, Mr Palomar sees the sun's reflection as a shining sword in the water. He swims towards it, but the sword retreats with every stroke and he is never able to overtake it. Wherever he moves he remains at the sword's tip. It follows him, 'pointing him out like the hand of a watch whose pivot is the sun'. He realizes that every bather experiences the same effects of the light. Sailboards change their appearance as they cross the reflection, colours are muted, bodies silhouetted. What if all the swimmers and sailboarders return to the shore, he wonders, where would the sword end?

Mr Palomar understands that nothing he sees exists in nature. Nature is a bundle of abstractions – particles in fields of force. The sun, the sea, the sword, and the sailboarders are inside his head. He floats among phantoms.

The sword of the sun cleaves the universe in two: there is objective reality – remote abstractions without point of view – and there is Palomar's private universe, the mirage of human perception. 'I am swimming in my mind; this sword of light exists only there.' But what kind of thing is Mr Palomar, the Perceiver? No doubt he would see himself, as I see myself, as a singular, unified being, continuous with his child self as I am continuous with the boy diving for his father's eye, moving from fixed past to uncertain future. Like the sun's reflection, this is an illusion.

In the angiography suite, performing a Wada test, I was the illusionist. Our brain drug cleft the young woman in two. The left-brain version of Naomi was different from the right. Ms Left-brain was talkative and cheerful. Ms Right-brain was unsettled, mute, morose. When the words finally broke through, she hadn't a clue where she was. 'What the fuck am I doing here?' I've never heard Ms Left-brain swear. Afterwards, when the drug wore off, Ms Left-brain spoke for the whole person. 'It was a breeze,' she said. There was no recollection of Ms Right-brain's discomfort. It had been edited out of the story.

One might think that the self is divided in such circumstances, but this would be to swallow the illusion of unity; to imagine in the first place that there is some 'whole thing' to be fractionated. There isn't. From a neuroscience perspective we are all divided and discontinuous. The mental processes underlying our sense of self – feelings, thoughts, memories – are scattered through different zones of the brain. There is no special point of convergence. No cockpit of the soul. No soul-pilot. They come together in a work of fiction. A human being is a story-telling machine. The self is a story.

This is not to say that our lives are fictions. Unlike Robinson Crusoe or Emma Bovary we are embedded in a universe with physical and moral dimensions where every thought and action splinters into a million consequences. Readers of Flaubert's *Madame Bovary* will vary in their reactions to its heroine as she makes her way through the novel, but her life and thoughts are fixed. She will always marry Charles, fall prey to the abominable Rodolphe, and die her horrible death. It's different for us meat puppets. We don't know where our lives are going. *What the fuck am I doing here?* I often wonder.

Who tells the story of the self? That's like asking who thunders the thunder or rains the rain. It is not so much a question of us telling the story as the story telling us.

Not so long ago I asked my dad if he remembered the time I rescued his eye from the bottom of the sea.

'No,' he said.

Soul in a Bucket

I once met a young man who was convinced his head was full of water and contained a fish rather than a brain. It was quite a large fish, something like a trout, and it unsettled him to think of it living in such cramped conditions. He no longer had need of a brain since all his thoughts and behaviour were under the control of the CIA.

Most of us believe that the head contains a person: a self. Here's one, at the front of a lecture hall, spilling words that seem to come from nowhere. There, in the auditorium, are 200 other selves. Rows of heads. For each head to represent the location of a conscious self requires a further, inferential, step, a mental process I am powerless to resist.

The same applies when we reflect on our own identity. We create our selves by inference: automatically and irresistibly. In doing so we ride the rails of the deepest human convention, but, at root, it is just that: a convention. The self is not an intrinsic feature of the brain and it is possible to become derailed – through psychosis, like the man with the fish in his head, or as a result of brain damage. The degradation of personality is a neurological commonplace.

Mary had suffered a brain haemorrhage; to be precise, a ruptured anterior communicating artery aneurysm. The arterial wall had always been defective (though she was not to know) and now, in her fiftieth year, the sac had burst, pouring blood into the frontal lobes. The surgeons opened up her head and fixed a clip to stem the flow. She had been close to death. Three weeks later, sitting in my office, it was difficult to stem the flow of words.

'I've got a poem I wrote it yesterday well I haven't written it down it just came to me when I was sitting looking out the window at the lawn and these magpies came vicious things you wouldn't want to leave a baby outside they'd peck its eyes out like they do the sheep they attack in pairs they swoop down and confuse the sheep one then the other we had a kitten climbed an apple tree it did they flapped around her poor thing was terrified I threw a stone we shook the biscuit box to get her down.'

She paused. She had forgotten the poem.

'Where was I?' she said.

I didn't say a word as I reached for the black case containing my test equipment, and avoided eye contact. If I didn't speak and I didn't look, Mary would stay silent. Without the trigger of a word or a glance she sat, if not exactly still (she was always fidgeting with the buttons on her blouse) then at least quiet. But to let slip a careless word or glance was to open the sluice. I quickly released the clasps on the case and took some clean history sheets from a tray on my desk. Mary didn't move a muscle. She was not even fiddling with her buttons. I wondered how long we could sit there like this, motionless and quiet. The silence didn't trouble her.

She seemed absorbed by a picture postcard pinned to the board behind my desk. It showed a Mediterranean scene, a seaside town with a pine-fringed golden beach and blue sea, and a seafront promenade with colourful shops and restaurants. MAJORCA blazed diagonally, upper left, in curly yellow letters. It had been there since the summer and now looked incongruous beneath the seasonal tinsel and plastic holly my secretary had stuck about the place. A little Christmas tree sat on the filing cabinet in a clutter of cards.

We began with questions about orientation. Time, place, and person: the when, where, and who co-ordinates of personal awareness. It's important to exercise discretion. One doesn't want to insult the patient by asking overly simplistic questions. But, with Mary, it was appropriate to start with the basics. Personal orientation was one of her problems.

'What day is it?'

'Wednesday.'

'Good. And the date?'

'Is it the twenty-fourth?'

'Actually it's the sixteenth. What month is it?'

'July.'

'What makes you think it's July?'

'It's warm in here.' She undoes a button, then another.

'I should keep your blouse on, Mary,' I tell her. We move on. 'Where are we now?'

'At the hotel.'

'And what is the name of the town we are in?'

'I don't know,' she says. 'Majorca, somewhere.'

I ask her name. She gives me a pitying look.

'Me? I'm Mary Magpie. Who did you think I was?'

* * *

I've been projecting images of the brain on to a large screen. At first they were hyper-real 3-D images, labelled and colour-coded to illustrate the anatomical landmarks. The cerebral hemispheres looked like they were made of shiny plastic. But now I am working on a more schematic picture. A large block of colour – hot mustard yellow – slides down behind me, casting a *trompe l'œil* shadow against the pale screen. It bears the legend CEREBRAL CORTEX and signifies the apparatus of the conscious self.

The hall is full and the students are attentive. They seem to have enjoyed watching the shapes and words glide across the screen, falling into place with *PowerPoint* precision as the brain assembles itself. I am pleased with my picture. It is like a work of art. The lecture hall as gallery.

In the shadows, on a table next to the illuminated screen there is another exhibit. It is hidden from view, but in a few minutes I shall take it from its container and hold it aloft for the audience to admire. For now, we contemplate the diagram. It provides a standard representation of the major anatomical divisions (hindbrain, midbrain, forebrain) and some of the component structures (cerebellum, thalamus, basal ganglia, neocortex).

This is an introductory lecture. I keep it simple, but move swiftly through the lower structures like a child ascending a climbing frame, eager to reach the top. I am most interested in what goes on in the higher reaches, the zones containing the interlinked systems of perception and thought, memory and emotion. Consequently, my account of the hindbrain and midbrain structures is crisp. I encourage the students to imagine that we are crawling through the base of a gargantuan skull and clambering up the brainstem. It has the girth of an oak. We proceed under the shadow of the great lobes of the cerebral hemispheres that loom like thunderclouds. It's the higher branches we aspire to, way up in the gloom. I ask them what they think they would see.

'Nothing,' one of them answers correctly, 'it's pitch black.'

'Shine a light,' I say.

'How old are you, Mary?'

'Twenty-four.'

'And your children, how old are they?'

'Emma's twenty-two, Tom's nineteen.'

I explained to Mary that we were in a hospital. Did she know why she was there? Yes, she told me. She'd had an aneurysm, but they'd put it right. She would be going home soon. In fact she would leave as soon as we had finished. It was a pity her sister had to stay behind. Her sister? Yes, she'd had an aneurysm too, but she wasn't doing so well. She would have to stay on the ward a little longer.

Mary was becoming agitated now. She stood up and made for the door.

'Got to go,' she said. 'I left the baby.'

'What baby?'

'My baby,' she said. 'I left it in the garden. Those magpies will have its eyes out.'

The baby was born last month. They opened Mary's head, and then they delivered her baby. It was a beautiful little girl, but there was a problem with her brain. It could be an aneurysm. Perhaps they'll open her head as well.

Shine that light at the glossy underside of the temporal lobe, directly above, and you will see that the outer surface is wrapped in a sheet woven from an exquisite material. Next slide: This is the 'grey matter'. It covers all of the major lobes. In reality the colour would be a dull, grey-brown, but here we'll give it a silver sheen. Dissolve into the fibres of this material. Picture an exotic, illuminated garden. See what makes it glisten. The objects all around ('neurons') are certainly plant-like – roughly spherical pods with slender, branching tendrils (the 'dendrites') and a longer process (the 'axon') extending from one end. The axons, too, can be seen to branch into a number of finer strands, each with a button-shaped endfoot that attaches to a dendrite, or to the cell body, of another neuron.

It is a dense matrix of interconnection. Above and below, near and far, the neurons pulse and glow in a silent, iridescent fugue as electrochemical signals traverse the long axons and influence the target cell. Next slide: Here, for our benefit, packets of light shoot along the axons and cause the cells to which they are linked to glow either red or blue. Red signals a state of excitement. The target cell itself is now encouraged to fire pulses along its own axon to cells further along in the network. Blue signals inhibition and the target cell stops firing. In effect, each neuron is either 'on' or 'off', generating pulses or ceasing to generate pulses. Neurons are the basic functional units of the brain and that is their task: to fire or not to fire. It's all they do. Whichever region of the cortex you plunge into, the scene is the same.

Where is the mind in this tangled wood of neurons and nerve fibres? It isn't anywhere. And the self? What did you expect? A genie in a bottle?

Gottfried Leibniz, the eighteenth-century philosopher and mathematician, performed a similar thought experiment. He imagined 'a machine whose construction would enable it to think, to sense, and to have perception' and, further, that the machine is 'enlarged while retaining the same proportions, so that one could enter into it, just like into a windmill'. What does he find in the interior of the mind-making machine? '. . . only parts pushing one another, and never anything by which to explain a perception'.

The enigma of personal identity may have a dark side. In his essay 'Sorry, but Your Soul Just Died' Tom Wolfe imagines an apocalyptic near future where advanced methods of brain imaging will strip away the illusion of self. People will realize that all they are looking at is a piece of machinery, devoid of self, mind, or soul. At this point, he says, 'some new Nietzsche' will step forward to announce the death of the soul and 'the lurid carnival that will ensue may make the phrase "the total eclipse of all values" seem tame.'

It is true. Neuroscience is fast developing the technical and conceptual wherewithal to reveal in fine, bare detail the neurobiological substrates of the mind. Perhaps it will despoil a sacred myth – the myth of selfhood and souls. And, if so, we may be wandering innocently into the opening phase of a dangerous game. Our ethics and systems of justice, our entire moral order, are founded on the notion of society as a collective of individual selves – autonomous, introspective, accountable agents. If this self-reflective, moral agent is revealed to be illusory, what then?

Values may have more to do with primitive ideas about *ghosts in machines* than we care to think, and perhaps by using the tools of neuroscience to deconstruct the self we run the risk of splitting a social atom and releasing forces beyond our present comprehension. Could 'the century of neuroscience' really signify the death of the self and the collapse of all values? I think Wolfe honours neuroscience unduly. He is seduced by the gadgetry and the gaudy images. You don't need futuristic new technologies to expose the brute fact that there's nothing but meat inside our heads. We've known this down the ages.

It dawned on me some time ago that I was no longer especially interested in the brain. Or rather, that my interest was expanding outwards from the brain itself. It was as if I had been in a congested city, gawping at the crowds and the architecture and the traffic from ground level. And now I was rising above the buildings and the streets to take in a different perspective. I could see suburbs and fields and rivers beyond and, in the distance, other towns and cities. Cities don't float in a vacuum, and neither do brains.

What became clear was that the brain could not be fully understood if you treated it as an isolated object. I had underestimated how tightly the brain's functions are bound to the rest of the body and, at the same time, how deeply they are embedded in the wider physical and social landscape. No brain is an island.

When Mary's husband came to visit he had a calming effect. They seemed to function as a unit. Mary's behaviour meshed into the networks of partnership and so became more coherent and consistent. In any relationship each person is partly defined in terms of the other. So, for Mary, her husband's presence was a guide to self-definition. He provided a template. He drew from her a behavioural repertoire and a mental structure to complement his own, and the centre of gravity lay between them. There was stability, a kind of equilibrium. This effect was not of his deliberate doing. That's just the way it happens.

If Mary's heart or lungs or liver had been the primary site of pathology, rather than her brain, it would be possible to describe the disease in terms of its effects on that particular organ system in relation to the overall functioning of the rest of her body. The function of the heart is to pump blood, the liver secretes bile, the lungs enable the supply of oxygen to the blood, and in each case the frame of reference for a description of function is the individual organism. In defining brain function we have to go beyond this, extending the frame of reference beyond the systems of the body.

The brain evolved as a means of orchestrating adaptive interaction between the organism and the world. To achieve this it must maintain both an inward and an outward orientation, monitoring and regulating the state of various internal systems, while at the same time responding to the flow of events in the external world. In fact, as well as playing its part in monitoring the body's internal milieu, the brain must control interactions with two kinds of external environment.

According to Western intellectual tradition, which distinguishes between Nature and Culture, we have a curious, duplex kind of existence. We move in a natural realm of time, space, and matter and, concurrently, through a socio-cultural dimension of people and ideas, a world saturated with customs and beliefs, rituals, traditions, laws, conventions, fashions, language, arts, and science. In the first world we are subject, ultimately, to the laws of physics and, in the second, to the influence of customs, beliefs, rituals, traditions, etc.

An emerging theme in neuropsychology is that, just as it has functional systems devoted to perception of, and interaction with, the physical environment, so the brain has evolved systems dedicated to social cognition and action. It constructs a model of the organism of which it is a part and, beyond this, a representation of that organism's place in relation to other, similar, organisms: people. As part of this process it assembles a 'self', which can be thought of as the device we humans employ as a means of negotiating the social environment.

Tightly bound to language, these brain mechanisms are the channels through which biology finds expression as culture, a means of distributing mind beyond biological boundaries. But if culture is in this way an extension of biology, an important question arises: must we also accept that neuroscience has boundaries which deny full access to an understanding of brain function? In other words, is neuroscience adequate to its primary task – understanding the brain – or, to tackle the 'big' questions (relating to self-awareness and personal identity) must we turn to other forms of science and scholarship?

To achieve some understanding of Mary's condition we are

obliged to skirt these fuzzy boundaries of biology and society. Beyond accounting for her illness in terms of physical pathology and appreciating its consequences at the personal level, we must try to understand what mechanisms might be operating at the intersection of the biological (the brain) and the social (the self). A major challenge for neuroscience in the twenty-first century will be to try to figure out how brains and selves go together.

We build a story of ourselves from the raw materials of language, memory, and experience. The idea of the 'narrative self' has a long history, with roots in Buddhist teaching. According to the doctrine of *Anattavada*, the self is no more than the aggregate of an individual's thoughts, feelings, perceptions and actions. There is no central core or 'ego'. David Hume, in the eighteenth century, took a strikingly similar line. For him, the extension of the self beyond such momentary impressions was a fiction. Daniel Dennett has offered a contemporary version, emphasizing the power of language in giving coherence to our experience over extended periods. According to Dennett, the self is best understood as an abstract 'centre of narrative gravity'.

Confabulation is the inadvertent construction of an erroneous self-story, signifying the neurological breakdown of the storyteller. It takes different forms, sometimes mundane, sometimes fantastic. As in Mary's case, it is typically associated with damage to the frontal lobes and is probably due to a combination of things. Memory disorder is one ingredient. In particular, confabulators have problems with contextual memory. They may retain the kernel of some autobiographical event or episode, but fail to anchor it in a specific time or place. Memories drift loose, images collide.

Then there is disinhibition of associations. Words, thoughts, and memories reach consciousness through a process of natural selection. For every item of awareness there is a multitude of suppressed alternatives reverberating through the neural nets. The confabulator's theatre of consciousness is crowded with gatecrashers (imaginary babies, magpies, the number twenty-four). The reduplication of relatives or the creation of imaginary children is a common theme.

Finally, there is a disturbance of the neuropsychological mechanisms responsible for maintaining a distinction between the external world and internally generated thoughts and actions (falling into the frame of a seaside postcard, you are transported to the island of Majorca).

I reach into the shadows behind the screen and retrieve a small, semi-transparent plastic bucket. I dip into the bucket and fish out a human brain. I have no idea to whom the brain belonged. I can't tell if it's male or female, black or white, or, with any reliability, its age. I may even have passed this person on the street. In its natural state, encased within the skull, brain matter is gelatinous. This brain, fixed in formalin, has a solid, rubbery feel and would carve like a very tender tuna steak.

It looks small and lacklustre after the bright pictures on the screen, but it holds the interest of my audience. The end-of-lecture rustling of papers stops. All eyes turn towards the grey-brown object as I point out the major landmarks. In terms of imparting factual knowledge about the structure and functions of the brain, the main purpose of my lecture, this little coda adds nothing. Yet the students leave with something they wouldn't otherwise have had: a clearer sense of the brain as a biological object; a physical mass as well as a textbook concoction of colours and neat abstractions. It will help them appreciate the distinction between the brain and the self.

When the Apollo astronauts went to the moon and brought back pictures of our planet of oceans and clouds hanging over a grey moonscape in the middle of a black nowhere, it changed the way we saw ourselves. We knew already that we inhabited the surface of a small, spinning sphere that rolled around an ordinary star, at the edge of an unremarkable galaxy, just one of indeterminate billions in a vast, indifferent cosmos. But now, occupying a few degrees of retinal space, comfortably absorbed in the folds of the visual cortex, a mere portion of the visual field, we saw our home in its true colours. It was precious and vulnerable, a small fragile object, a thing we should take care of. It was, indeed, our home. We might have extrapolated these sentiments from the knowledge we already possessed, but the images set off an interplay of intellect and imagination that made the new perspective irresistible.

Something similar happens when you see a brain. Imagination infiltrates intellect. You get a sense of location and vulnerability. Our home.

The hall empties, but a few students stay behind for a closer look. They want to touch it. A young woman asks if she might hold the brain. She dons rubber gloves and takes the specimen in her cupped hands. There is wonder and apprehension on her face. You see this look on the faces of small children holding caterpillars. A young man turns the brain over to examine its underside. He picks at the stump of a severed artery. Another tests the weight, feeling the drop of the object first in the left hand then the right. He says it's odd, but your head doesn't feel this heavy.

Six months later Mary came to the outpatients' clinic. It was a routine follow-up. She did not remember me but, although her memory was still poor, she had made good progress in other areas. There was no longer the prolixity of speech or the fragmented attention that had characterized her behaviour before. In particular, over a period of two hours of interviewing and neuropsychological testing, I detected no signs of confabulation. Then, work done, idly chatting as we waited for her husband to collect her, I asked what plans she had for the weekend.

'Oh,' she said. 'I'd like to watch the badgers again.'

'Really?'

'Yes, in the field over the back wall. You can see them from the garden shed.' She was looking at my name-tag. 'Brock the badger,' she said.

Mary's husband arrived and they left together. I never saw them again and I didn't ask him about the badgers at the bottom of the garden.

Like the surface of the Earth, the brain is pretty much mapped. There are no secret compartments inaccessible to the surgeon's knife or the magnetic gaze of the brain scanner; no mysterious humours pervading the cerebral ventricles, no soul in the pineal gland, no vital spark, no spirits in the tangled wood. There is nothing you can't touch or squeeze, weigh and measure, as we might the physical properties of other objects. So you will search in vain for any semblance of a self within the structures of the brain: there is no ghost in the machine. It is time to grow up and accept this fact. But, somehow, we are the product of the operation of this machinery and its progress through the physical and social world.

Minds emerge from process and interaction, not substance. In a sense, we inhabit the spaces between things. We subsist in emptiness. A beautiful, liberating, thought and nothing to be afraid of. The notion of a tethered soul is crude by comparison. Shine a light, it's obvious.

In the Theatre

In the days before brain scans it was impossible to locate tumours beneath the surface of the brain with any precision. Surgeons blindly poked around in the soft tissues, causing untold damage in the process.

The poet-physician Dannie Abse wrote a poem, 'In the Theatre', in which he recounts a harrowing experience described by his father, Dr Wilfred Abse. I first read this poem when I was a student and it raised the hairs on the back of my neck. It concerns an episode that occurred in 1918 while Abse senior was assisting at a brain operation.

His voice introduces the poem. The patient, he tells us, is fully awake throughout the operation under a local anaesthetic, while the fingers of Lambert Rogers, the surgeon – 'rash as a blind man's' – crudely search for the tumour in the brain tissues – 'all somewhat hit and miss'. This, says Dr Abse, is one operation he will never forget. The poem, and the operation, opens with reassuring words for the patient:

Sister saying – 'Soon you'll be back on the ward,'
sister thinking – 'Only two more on the list,'
the patient saying – 'Thank you, I feel fine';
small voices, small lies, nothing untoward,
though, soon, he would blink again and again
because of the fingers of Lambert Rogers,
rash as a blind man's, inside his soft brain.

If items of horror can make a man laugh
then laugh at this: one hour later, the growth
still undiscovered, ticking its own wild time;
more brain mashed because of the probe's braille path;
Lambert Rogers desperate, fingering still;
his dresser thinking, 'Christ! Two more on the list,
a cisternal puncture and a neural cyst.'

Then, suddenly, the cracked record in the brain,
a ventriloquist voice that cried, 'You sod,
leave my soul alone, leave my soul alone,' -
the patient's dummy lips moving to that refrain,
the patient's eyes too wide. And, shocked,
Lambert Rogers drawing out the probe
with nurses, students, sister, petrified.

'Leave my soul alone, leave my soul alone,'
that voice so arctic and that cry so odd
had nowhere else to go – till the antique
gramophone wound down and the words began
to blur and slow, '... leave ... my ... soul ... alone ...'
to cease at last when something other died.
And silence matched the silence under snow.

The poem no longer chills me. Why? Perhaps I am a more sophisticated reader of poetry now and see too clearly the boards and backdrop of a melodrama. But I don't think so. It is obviously meant to be theatrical and, to my mind, conveys authentic drama. It is a fine poem. It still packs a punch. But it doesn't unsettle me as it once did.

Perhaps my experiences as a clinician over the years have left me desensitized to human suffering or lacking appetite for the bizarre and extraordinary. I hope not, and don't think so. Inevitably, one develops strategies for self-preservation. Anyone who has worked with patients on acute hospital wards will tell you that you cannot resonate with every tremor of feeling, and that sometimes there are visions of horror and raw fear that can only be observed obliquely.

Perfect, constant empathy in such circumstances would be suicidal. But it is less a process of desensitization than one of becoming acclimatized. There is a difference. The former suggests an atrophy of feeling, the latter is merely to become accustomed to different conditions. When the professional façade slips in the presence of personal suffering, as it does from time to time, the pain still penetrates. And as for indifference to the bizarre and extraordinary, the very opposite is true. From where I stand, in early middle age, the universe looks as mysterious and absurd as ever – human beings especially. The older I get the more astonished I am by the plain fact of my own existence and consciousness, let alone strange and disturbing neurological cases.

The reason the poem has lost its power to give me goose pimples comes down, I think, to that word 'soul'. For full effect it requires an acceptance, at some level, that souls exist. Consider the pivotal moment when, defenceless against the surgeon's frantic fingers, the dying brain asserts itself: *Leave my soul alone, leave my soul alone.* This feels like the intrusion of a supernatural force. Something other than the brain, and something other than the patient even, seems to have entered the scene. It is hard to identify the voice with the soft mass of the inert brain, and it is distanced from the patient, whose body now becomes a ventriloquist's dummy, mouthing words as if to the sound of a cracked record. They are words of despair. The eerie intruder was reluctant to manifest itself, one feels, but had no choice.

Now, if you had asked me twenty-odd years ago whether I thought there were such things as souls I would have said that of course there were not. Undergoing training in clinical and scientific disciplines, I would have considered such talk to be primitive and pre-scientific. 'Soul' implied a spiritual substance of some kind, a mental essence or ego acting behind the material scenes of brain structure and function, guiding and controlling and, if you chose to believe it, surviving the death of the body. It carried connotations of the supernatural, representing ideas that I found at best misguided or intellectually inelegant, at worst sinister and retrogressive.

I took the view that the mind was the product of the brain in its interactions with the physical and social world. I still hold that view. The difference in my emotional response to the poem then and now requires a more subtle explanation.

I think it has to do with a change in background intuitions, as opposed to foreground beliefs. When I first encountered the poem I explicitly denied the existence of souls, but there was a part of me still beguiled by the imaginative power of the term. It feels natural to believe that something like a soul exists. It may indeed *be* natural in the sense that evolution has shaped our cognitive architecture in ways that predispose us to believe in the separation of mind and brain.

We live in a social as well as a physical world, and negotiation of our complex social environment requires the attribution of mental states (feelings, beliefs, desires, intentions) to ourselves and others, perhaps inevitably inclining us to believe that the world contains two sorts of stuff, material and immaterial. Dualism may have deep evolutionary roots. We all feel that, as well as a brain, something else occupies the interior of our head and the heads of other people – an irreducible mental core, the origin of thoughts and actions. It is a primitive belief, but it is compelling.

It may follow that a belief in souls (or implicit belief or half-belief or quasi-belief) is a necessary condition for ordinary human interaction. We view ourselves as integrated mental entities with an agenda of intentions and actions, authors of our own destiny. In a moral universe, perhaps that is the only conceivable way to view other people too.

All I can say is that now, when the brain/soul/patient pleads *Leave my soul alone,* it seems to me less like the intrusion of a supernatural force. Nothing has entered the scene. There is still a desperation for survival, but this, I now see, emanates from a stark no man's land between the mutilated tissues of the brain, the sound waves of 'that voice so arctic', and the horrified response of the onlookers. There is horror still, but not supernatural horror.

Perhaps, over the years, my immersion in the rationalism of clinical science has rusted those machineries of imagination that generate belief in supernatural souls – or half-belief or quasi-belief. Perhaps this should worry me; I may have lost something. But I prefer to think that the screen between the two domains – scientific understanding and primitive imagination – has become more transparent, allowing a clearer view in both directions.

Certainly, I feel better able to cope with ambiguity. There is scientific rationality and there is imagination. Sometimes they coincide, and all people have elements of both. The poem betrays an ambiguity about the reality of souls. The fearful words crank out through the antique gramophone of the vocal apparatus, blurring and slowing, '. . . *leave . . . my . . . soul . . . alone . . .*' – an image of a soulless mechanism – but they are said to cease at last only '*when something other died*'.

So, it doesn't chill me as it used to, but I feel another response. Behind the horror I see more clearly now there is also pity. The brain thinks it's a soul. There is real pathos in that.

I may not believe in souls, but I still find brain surgery disconcerting, and this is partly why Abse's poem retains some of its original power. The experience of witnessing an operation on the brain is, in certain respects, quite different from other kinds of surgery. All surgical procedures are invasive, but neurosurgery seems like the ultimate intrusion. In abdominal and cardiothoracic operations you see the surgeon burrow into the patient's body, exposing the inner workings – the pumps, the filters, the pipes, the valves. It can be shocking, but we have this way of separating ideas of 'the person' from what is happening to 'the person's body'. The person, one tends to imagine, is elsewhere during the invasion of her innards; she has withdrawn to a safe, anaesthetized place (somewhere in the head, our intuitions tell us).

It is a different matter when the contents of the skull are open for inspection and prey to the surgeon's knife. There can be no question of any 'relocation' to another part of the body. It would be absurd to think of the ghostly self retreating to some other organ or limb. At the same instant one understands that there is, of course, no ghostly self in the first place. When we see the brain we realize that we are, at one level, no more than meat; and, on another, no more than fiction.

The same insight – at once mundane and mysterious – is a potent element of the famous Zapruder footage of the Kennedy assassination. Recall the images and you will know what I mean. Jackie Kennedy in the Dallas sunshine, in her pretty pink, pillbox hat, scrambling to retrieve the debris of her husband's shattered head from the back of the limousine. She finds a piece of something and tries to replace it. Or so it appears. What grim desperation; what appalling intimacy.

Those glimpses of naked brain are an epiphany. The Most Powerful Man in the World, smiling immortally and waving to the crowds just now, has had the top of his head blown away. It is not just that he, like all of us, is vulnerable, or that his physicality is so brittle. Nor is it even the suddenness of his transition from human being to inert substance. What locks these images into place is the exposure of the grey-pink brain. If that is what The Most Powerful Man in the World can be reduced to, if that is what he *is*, then there is no hope for the rest of us. We knew it anyway, but our awareness of the fact is intensified by the drama and the mythic status of the actors. That is the horror of the film. And when you watch, do you not sense a flicker of self-pity? Despite myself, I fear for my soul.

A–Z

A former neurosurgeon colleague, recalling his training days, has a story about walking through the streets of London with a fellow student of surgery. His friend stopped, mid-conversation, evidently troubled. They had reached a small side street. The man was becoming agitated and kept looking up at the street name.

'This street,' he said, 'is not in the A-Z.'

And it wasn't. They checked. The would-be surgeon had a remarkable visual memory. For no particular reason he had set himself the task, successfully accomplished, of memorizing the entire A-Z street map of London. And now he had stumbled upon a mismatch. The map of his imagination would have to be amended in one small detail, enough to make it superior to the published version.

His problem though (and ultimately what barred his way to a career in neurosurgery) was that his exceptional powers of visual imagery were exceptional only in two dimensions. Among other attributes of intellect and temperament, the practice of neurosurgery demands an ability to think in three dimensions.

This man was like Mr Square, the lowly, two-dimensional character in Edwin Abbot's nineteenth-century satirical tale *Flatland: A Romance of Many Dimensions*. Mr Square has no inkling of a geometry beyond the plane of Flatland until, one night, he is visited by Lord Sphere, a being from the land of three dimensions (appearing to him as a circle, magically changing shape). Failing to convince by explanation, Lord Sphere peels his humble acquaintance from Flatland and flings him into Spaceland, proving that there is, indeed, a world of three-dimensional objects. It is a revelatory, life-transforming experience. But, of course, on his return, he fails miserably to persuade his fellow Flatlanders of the existence of this other world and the High Priests (who are circular) condemn him as a dangerous heretic.

The trainee surgeon's A–Z of the brain lacked the necessary third dimension. He wasn't able to inhabit the metropolis of the brain in the way a neurosurgeon must.

Neuropsychology requires four dimensions. At least.

The Mirror

Judy gets home from work. She pours herself a martini, puts a record on the hi-fi and drops into a soft leather chair. *God* she's tired ... She sleeps. Andy can read the bedtime stories. The little girl kisses her sleeping mother and, reluctantly, goes to bed.

When Judy wakes, the room is semi-dark. She sits bolt upright and flinches from the pain in her head. There is no music. She tries to fix her position. Images of the daily routine tumble together: get up, go to work, come home. A cat stretches on the back of a chair by the window. There is enough of the fading light to bring its ginger fur to life. But Judy's cat is a tabby.

An unfamiliar man enters the room: late middle-aged, grey hair. He glides by, glancing briefly in her direction, and switches on a lamp in the corner. It casts a cone of light upwards to the ceiling.

'Did you say something, Jude?'

Now that the light is on she sees that the room is also unfamiliar.

Where am I? Who is this?

Judy has a habit of twisting and pulling at her wedding ring when she's stressed. The ring has gone.

'Jude,' the man says, 'I don't know what you're on about.' She is asking for her daughter and husband. The man leaves and a woman enters. She kneels by the side of the chair and takes Judy's hand. It's a gentle interrogation.

'I'm Judy Jenkins. I'm thirty-nine – and I don't know where the bloody hell I am!'

But she knows the year: 1976.

The Prime Minister? 'Harold Wilson.'

The man's voice breaks in, but Judy cuts him short.

'Don't tell me I'm getting mixed up!' she snaps. 'I know what year it is.'

Now he's in front of her holding up a newspaper, pointing to the date at the top of the page: Saturday, 10 April 1999. A shaft of logic breaks the frame.

'Fetch me a mirror.'

* * *

1999

'And then I said, "Fetch me a mirror." I said I'd be an old woman in 1999 . . .' She's telling me the story again. It must be the fourth or fifth time. I'm not really listening. I don't need to. It's the same story.

The neuroscientist Michael Gazzaniga quotes John Updike and Ralph Waldo Emerson: 'A thread runs through all things: all worlds are strung on it, as beads: and men, and events, and life come to us, only because of that thread.' In other words (says Updike) our subjectivity dominates outer reality, 'and the universe has a personal structure'.

According to Gazzaniga, the brain has a dedicated system for binding the strands of a multitude of specialized brain modules into a single thread of personal experience. He calls it 'the Interpreter' and locates it in the left cerebral hemisphere. The Interpreter identifies patterns of connection between disparate brain systems and correlates these with events in the external world. This gives unity and continuity and enables each of us to create a personal life story.

Judy's mirror arrives. She sees that her face is wrinkled and gaunt, her hair short and grey. *It must be a dream*, she thinks. It's too much to take in all at once. She panics and tries to stand, but her left leg is paralysed by the stroke. She loses consciousness.

When she wakes there is a man leaning over her, pressing a mask to her face; a soldier perhaps, though his uniform is too gaudy for the military. She senses movement and vibration and recognizes the sounds of an engine going through the gears. She is in some kind of vehicle.

'Take it easy, Judy,' says the man in uniform, 'you're going to be just fine.'

There are red and grey blankets and cylinders, tubes and leads, plastic boxes, instruments with dials, and other items of equipment. And there, sitting opposite, is the grey-haired man.

He leans in closer to her. 'What's that?' he asks.

'Has anyone told Andy?' she mumbles.

The erasure of twenty-three years is remarkable enough. Judy's personal life over that period is a complete blank. She can only comprehend the bitterness of her divorce from Andy in the abstract, and the grey-haired man, with whom she has been living for the past eighteen years, is a stranger. The same goes for public events. ('Margaret Thatcher?') But equally striking is the work of Judy's Interpreter, her teller of tales. Its struggle for continuity is heroic. With nothing else to hand, it reaches back more than two decades to find material for a story. It bridges the twenty-three years as if they were twenty-three minutes.

* * *

2002

'And then I said, "Fetch me a mirror" . . .'

Beaten into submission by the logic of an unknown present, plus the irrefutable evidence in the mirror, Judy's Interpreter has reset the clock and set about the business of regulating a different life. Very little memory has returned, but things are going fine with the grey-haired man, and Judy is a grandmother now.

This kind of amnesia is extremely rare. I have come across only one other case that bears comparison to Judy's. That patient, too, made an astonishingly smooth adjustment to her new circumstances. How resilient people are. If one day you woke up to find that you had been transformed into a gigantic insect, the chances are you would just get up and carry on with your new life.

* * *

2003

'And then I said, "Fetch me a mirror" . . .'

The Visible Man

As James Moon awoke one morning after disturbing dreams, he found himself transformed into an anatomical illustration. He rose from his bed thinking all was as it should be and headed for the bathroom, the first routine act of another routine day. But, looking in the mirror, he noticed that the dome of his head had become transparent.

There, bathed in a glossy light, was his brain, looking like a vivid picture from the pages of a textbook or a high-resolution computer graphic. The outer surfaces, the great convoluted lobes of the cerebral hemispheres, were rendered in pastel shades with sculptural clarity: compact, rounded, solid, with sharp contours defining the major anatomical divisions.

The frontal lobes (pale mauve) were packed just behind the forehead. The temporal lobes (powder blue) were to either side at the level of the ears. Above and somewhat behind these were the parietal lobes (champagne) and, at the back, the occipital lobes (jade). Then, as you would expect, there were cutaway views from the side, from above, and face-on, revealing structures deep beneath the surface. These were coded with bolder colours: cobalt blue, lemon, cherry red, orange, purple.

The closer he looked, the more James saw. Not just the bulbous, fruit-like forms of the putamen and the globus pallidus, and the flat-topped oval of the thalamus at the centre, or the sweeping, overarching curves of the fornix and the caudate. But there, like bright clusters of jellybeans, were the subcomponents of these structures: the pulvinar, the lateral geniculate body, the dorsomedial nucleus. He did not yet know the names of these things, but he would.

Of course, it's a dream, he thought, *I'm still asleep.* But the mundane objects of the bathroom stood in their usual places and on the sill was a copy of yesterday's newspaper, just as he had left it, a picture of a woman smiling on the front page. Wind and rain beat against the window pane. He would seek advice without delay.

So, littering crumbs of breakfast toast as he went, James grabbed his hat (a green canvas one like anglers wear), slammed the front door behind him, and set off for the medical centre.

One corner of the waiting room was partitioned as a children's play area. It was strewn with soft toys and plastic bricks in primary colours. A small child sat in the middle, tugging a string from the back of a pink doll.

'I'm so happy,' said the doll. The child chuckled as the string slid back inside.

All the while Greg knew that, beneath the hat, his head was still glowing like the aurora borealis. *And wouldn't that amuse the child,* he thought. He resisted the temptation, however, and was soon called through.

'How can I help?' said Dr Vesalius.

James removed his hat. 'Well?' he asked. 'Have you ever seen anything like it?'

The multicoloured radiance seemed brighter than before. It shone like a halo. If Dr Vesalius was shocked he didn't show it. A true professional, he just leaned forward, pressing together the tips of his fingers.

'Yesterday all was well,' James explained. 'Then this morning I wake up to find my head transparent and my brain a firework display.'

'I see,' said Dr Vesalius. He said he would make arrangements for James to meet a specialist.

On the way out there was another child playing with the doll. She pulled grimly at the string. 'I'm so happy. I'm so happy . . .' the doll kept saying, though the child was not amused. James lifted his hat briefly, but it made no difference.

* * *

What had caused those nightmares? The previous evening, James had eaten a light supper and drunk no more than a finger of whisky. Bored with TV, he had searched for something to read and found an old illustrated encyclopaedia, which he took to bed.

Idly turning the pages, he came across a picture of a rock pool. He remembered it well from his childhood. The water was clear as glass. Blue-grey rocks thrust up through a bed of smooth pebbles and sand towards a summer sky. Above the water-line there were barnacles baking in the sunshine, a few limpets and whelks – snail-like things with curly shells.

Mussels clustered and tumbled through the filmy surface of the water and into the psychedelic world below, which was brimming with all kinds of life: crabs and shrimps, slugs and starfish, sea anemones, spindly prawns, and grim-faced little fishes darting through the bladderwrack and brown seaweed. It still fascinated him, still drew him into the scene, but not quite through the page and into the water as before.

Looking at the picture now, through adult eyes, was oddly disillusioning. He had often walked along the beach at low tide and had yet to find a rock pool so stuffed with life and colour. But it wasn't just disillusionment. There was something else equally unsettling. He closed the book and was soon asleep.

In one dream he found himself tightly bound, head to toe, scarcely able to breathe. Aware of a slight swaying motion, he had the sense that he was lodged high in the branches of a tree or suspended in a net of some kind. There was a nauseating jolt, and another, as if he were being dragged along like coal in a sack. One particularly violent jerk pulled the binding from his eyes and what he saw wrenched his gut.

What had appeared to be the shadow of a black awning turned out to be the belly of a monstrous spider. He tried to break free, but it was useless and he was soon embraced by the beast's slavering maw. There was no pain, just warmth and moisture. As the bonds broke he saw his own disintegrating body: his segmented brown belly, his six trembling legs, the membranes of his buckled wings. He called out, but his voice was a twittering squeak.

And now this. James stared, without expression, at the mirror, then sank into an armchair. He felt inordinately weary.

* * *

It was dark when the doorbell rang. A pastel glow followed James down the dingy hallway, reminding him of the state of his skull. He took the fisherman's hat from the coat-hook and placed it on his head before opening the door. There stood Millie, in a swirl of rain and autumn leaves.

She had brought with her two bags full of books, which she was now stacking up on the kitchen table.

'These are from the library,' she said, 'as requested.' Among others, there was a medical textbook, an atlas of neuroanatomy, and a massive tome called *Elements of Cognitive Neuroscience*. 'And this one I bought.' She handed over a slim paperback: *Neuroscience for the Brainless*.

They sat at opposite ends of the sofa, Millie with arms folded, James gripping the brim of his hat. She was looking away as she spoke: 'All right, keep it on.' Her cheeks were red. After unpacking the books she had chased James around the flat, snatching at his hat. It was a game at first, she thought, but then he shouted at her to leave him alone. They sat in silence.

'Okay,' he said, 'I'll take it off.'

The delicate spray of light danced about his head and he wondered why he should have been so coy with Millie. Why should she not see this most enchanted part of him, this magical wellspring – the source of his thoughts, his hopes and beliefs, and of his love for her?

'Well,' she said, 'what was all the fuss about?'

He might have been comparing fossil specimens or semiprecious stones. 'The colours are different,' he noted, 'but the shapes are the same.'

Elements of Cognitive Neuroscience was propped up on the kitchen table next to a shaving mirror. A picture of the brain filled most of one page. Millie stood behind him, her eyes roving from the picture to the face in the mirror to the top of James's head. He watched her reflection looking down on to the shimmering surface of his cerebrum, her eyes wide, transfixed, confused.

'I can see it's going to take me a while to find my bearings,' said James, staring at his hand, but Millie had already left the room.

Make a fist with fingers wrapped around thumb. This is the brain. Palm upwards, the outer ridge of the forearm becomes the *spinal cord*. It turns into the *brainstem* at the wrist. Now look at the fleshy part leading up to the base joint of the thumb. This is the *hindbrain*. The protruding base joint itself represents the *cerebellum*, which is the most prominent feature of the hindbrain. In reality it looks like a kind of vegetable outgrowth at the brain's rear underside.

Moving upwards and into the tunnel of fingers, the shaft of the lower thumb bone represents the top end of the brainstem. This is known as the *midbrain*. Finally, there is the *forebrain* – the upper thumb bone, hidden under the fingers, and the fingers themselves. Each finger stands for a division of the topmost part of the brain – the *cerebral cortex*. Starting with the index finger, we have the *occipital lobe* (ok-SIP-itul), the *parietal* (puh-RYE-etul), the *temporal* and the *frontal lobe*. The upper thumb bone represents various forebrain structures that lie beneath the cerebral cortex (the *amygdala* and *hippocampus*, for example).

There you have it. The gross anatomy of the brain – or half
of it. The brain is a double organ with two mirror image sides.
Put both fists together to get the full picture.

Bruno Aldaris, *Neuroscience for the Brainless*

Though it was plain to see that his brain was a physical mass,
like a hand or a foot, James found the comparison of brain and
fist mildly disconcerting and soon returned to observing the real
thing.

It was Millie's idea to go to the Chinese restaurant. James
wore a baseball cap.

'You need to get yourself out of yourself,' she said.

They drank white wine and James began to unwind. He even
squeezed her knee under the table.

'It's going to be all right,' he said on the way home. 'You'll
see.'

This being a Friday, Millie stayed the night, but was too tired
to make love.

* * *

The dawn light seemed to grow in steps as if the Earth's rotation
had developed a fault. James lay listening to the rain, dipping in
and out of sleep. Millie lay beside him. He watched her dreaming through the blind saccades of her lidded eyes. Her brain was
in darkness, but the dreams, no doubt, were as bright as day.

He must have drifted off again because, next, he was aware
of the smell of fresh coffee. Millie had been out for croissants
and newspapers. James didn't think his brain was a sight for

the breakfast table and remembered to put on his hat before going through to the kitchen, but Millie, through a mouthful of croissant, told him instantly to take it off.

She returned to her newspaper. He picked up the medical textbook, scanning the list of contents as if reading from a menu: *Dementia; Cerebrovascular Disease; Hydrocephalus; Epilepsy; Extrapyramidal Disease; Cerebral Tumours; Demyelinating Diseases; Diseases of the Spinal Cord; Motor Neurone Disease.* That's just Neurology. Not even half of Neurology. Then there's *Cardiovascular Disease; Endocrine Disease; Haematological Disease; Gastrointestinal Disease;* and *Cancer.*

He was impressed by the myriad forms of demise. God was an inventive destroyer as well as an artful creator. It had never occurred to James that the workings of the body could break down in so many ways. But his condition was nameless.

He had two mirrors now, using them in combination for the difficult side and rear views, and was trying to match the textbook diagrams with the polychromatic contours of the object filling his head.

'It's a beautiful machine,' he said aloud, though there was no one else around. Millie had left him to it. 'Or is it a place?'

The books differed in emphasis. Some were concerned with systems and functions, others, especially the atlas of neuroanatomy, concentrated more on the brain's geography, conjuring strange undulating landscapes. Combining the different images, James pictured a metropolis, at once futuristic (full of mysterious machines) and ancient (the Greek and Latin names evoked classical times). Seen this way, his brain became a labyrinthine structure with vaults and chambers, floors and screens, columns, pathways, bridges, canals and aqueducts, with streams of information flowing in every direction. *Am I in there or out here?* he wondered.

'Am I out here or in there?'

First the thought, then the words. Thought. Speech. Thought. Speech. Alternating between the two, eyes fixed on the image in the mirror, he noticed a pattern, an ebbing and flowing of activity on the outer surface of the left frontal lobe. As he spoke, the soft mauve luminescence seemed to harden momentarily into a brighter glaze that dissolved as the utterance stopped. Looking closer, he saw strands of light casting back and forth between the frontal area and the blue recesses of the temporal lobe. And was that a fainter pulse, deep down? *Probably the thalamus*, he thought, consulting the atlas.

Am I in there or out here? The more James stared into the mirror, the more perplexed he became. He began to feel detached from his brain – a remote observer. It acquired the aura of something alien, an object quite separate from him. He would concentrate on a patch of colour or listen to a sound or perform an action or think a thought – *Elephants are large mammals. Six sevens are forty-two. Democracy is a good thing. I love Millie* – and there, plain to see, were the correlated brain patterns.

But the activity associated with thoughts and actions was not the same as his conscious awareness of those thoughts and events. How could it be? He was looking in from the outside, like watching goldfish in a bowl. Goldfish and one's perception of them are not the same thing. The more he gazed upon it, the more James felt that he was something other than his brain.

And where do thoughts and feelings come from? *Not me*, he thought, because he saw that every fluctuation in the flow of experience, every intention and action, was *anticipated* by distinct tremors of activity across the brain's surface. It was not a case of thinking or doing something and watching the brain follow step or dance in synchrony. His brain was ahead of him. Ideas were bubbling up in the neuronal cauldron a good half-second before they appeared in consciousness, even thoughts about thinking thoughts, and thoughts about thoughts about thinking thoughts. So who was stirring the mental broth? And if he were a mere spectator, what exactly was his vantage point?

But then, just as a drawing of a cube seems to change perspective, continually jumping inwards and outwards, he would switch to a different view. The object in his head would absorb and beguile him, and he would identify more closely than ever with its workings. *That's it*, he thought. *That's what I amount to. This is my sum total. There is no one stirring the broth. The functions of the brain have a life and logic of their own. Thoughts, feelings, and intentions produce me, not the other way around.*

He came to the conclusion that he was neither *in there* nor *out here*. Both perspectives were false. He wasn't anywhere.

In the evening James and Millie went to the cinema. James kept his hat pulled firmly down. It was hardly the place to have one's head blazing like a beacon.

* * *

A vague curiosity drew him back to the encyclopaedia. There was something wrong with the rock pool, he was sure. The plants, pebbles, and fishes were displayed as treasured objects in a glass case, the exhibits separated evenly one from another in a perfectly illuminated three-dimensional space. The artistry was calculated. Those beautiful things were intended to be memorable. He knew he would have no difficulty recognizing a *rock goby* if ever he saw one, or a *cushion starlet* or a *chameleon shrimp*.

Yet, although the picture was crammed with information, he could appreciate what was lacking. There was no sense of process or behaviour, nothing of the struggle for existence in the world-between-tides. One did not see the monstrous dog whelk devouring the barnacles or boring into the shell of a mussel to suck out its innards, or the invisible alchemy of the seaweeds, absorbing sunshine, synthesizing foods from water and carbon dioxide, releasing life-sustaining oxygen into the water.

The scene was altogether too benign. A rock pool is, in reality, a precarious place. To survive is to mesh with complex networks of behaviour and intricate patterns of physics and chemistry, all shaped minutely by the ebb and flow of the tides and the rotation of the Earth. The life of the rock pool is a fragile product of the microscopically small and the astronomically large.

But that was not it. Illustrations only ever give a snapshot.

He reached for one of the shiny new textbooks and set it alongside the encyclopaedia. The brain pictures were brightly coloured like the rock pool and, in the same way, were concerned with categorical clarity: this little fishy is the *hippocampus*, this the *medulla*, this the *cerebellum*. The creatures floated in suspension beneath the rippling surface of the cortex. The flow of time had stopped. There was no hint of dynamics at the microscopic level or of forces in the world beyond the boundary of the skull, both of which shaped the activity of the brain, just as the life of the rock pool was shaped by photosynthesis and the gravitational influence of the moon.

It was then that an unsteadying thought swung into James's head. He watched it drop from the frontal cortex and circle the limbic lobe. A brain, like a rock pool, he realized, is also a most precarious habitat. The life of the self depends absolutely on the integrity of brain function.

And then he saw what it was about the rock pool that really troubled him. There was a creature he hadn't noticed before, a squat, spidery thing, dull-grey and ugly compared to the rest. It lay beneath a *brittle star*, part hidden by a curtain of *brown weed* and the claws of a *shore crab*. This little fellow didn't figure in the key. It didn't have a number. Perhaps it had crawled from behind a rock. Perhaps the picture had other dimensions after all.

He looked at the mirror, into his brain. And there it was, the spidery thing.

* * *

Dr Stroop's office on the fourteenth floor of the District General was a shambles. There were box files, case notes, books piled all over the place, and cardboard boxes full of other stuff. On top of the filing cabinet, caught in a shaft of late afternoon sunlight, was a plastic model of the brain, somewhat larger than life-size. Its colours seemed to fill the room and James felt uneasy sitting near to it.

'I find that most people try to ignore it,' he said, rolling his eyes upwards, 'but it must be of some professional interest to you.'

'Oh, what's that?' asked Dr Stroop.

'The colour coding is much the same as yours. Do you mind?' James lifted the plastic brain from its stand and ran his fingers over the surface. 'The frontal lobes are quite similar, and the parietal, but my temporal lobes are blue and these occipital lobes are a darker shade of green.'

He leaned forward for Dr Stroop to get a better view, but the doctor didn't seem very interested. Raising his eyes, James caught him looking at his watch. He would have asked him about the spidery thing – perhaps he would be able to identify it – but could see that he was well behind with his clinic and didn't want to cause further delay. Bringing the consultation to a close, Dr Stroop muttered something about 'diagnostic investigations' and said that arrangements would be made for a brain scan. *Hardly necessary in my case*, thought James.

The next time they met, Dr Stroop had a sombre look, but he spoke kindly.

'Do you have anyone with you?'

'No,' said James. The doctor asked him to sit down.

'We have your pictures,' he said with an uncertain smile. He slid a large square film under the clip of a wall-mounted light box and flicked the switch. The image of James's brain was monochrome and murky, except for one feature. Could that be the spider, sitting plump and bright in the middle? If so, it had been busy. Skeins of cobweb spread like the wings of an angel, one into the right hemisphere, and one into the left.

'What is it?' said James.

'A butterfly glioma.'

That sounded rather beautiful.

'A tumour.'

That evening as he splashed water on his face, James noticed something in the palm of his cupped hand, or rather something *about* it. It was the left hand. The skin was paler, like greaseproof paper. There were shapes beneath: tracings of tendons, bones, and blood vessels. He inspected the rest of his body. His arms and legs looked fine, as did his chest and abdomen. He turned his penis left and right and lifted it to see the underside. He carefully checked the soles of his feet. Everything seemed normal. Turning away from the mirror, he looked across his shoulder and saw that his lower back was translucent. Spinal column, pelvic bones, and ribs were clear to see. And parts of the internal organs: stomach, liver, intestines.

He was struck by how densely packed they were, not floating loose like odd-shaped balloons as they sometimes appear in textbooks. But the colours! That couldn't be right: ice-blue stomach, crimson liver, orange intestines, and the bones were white as a ghost-train skeleton's.

It didn't stop there. The following morning he sat once more with his mirrors and books. It was odd that he hadn't noticed the butterfly wings before, but having seen them on the brain scan picture and knowing exactly where to look, he could now trace their outline.

Butterfly glioma came under *Cerebral Tumours* in the medical textbook. The prognosis was poor, but how fortunate he was to have been spared the psychiatric symptoms that, according to the book, were often associated with such growths. His thoughts turned to Millie. An image of her face shone as bright as stained glass in his mind's eye. Where was that, exactly? *Neither in there nor out here.* And where exactly was Millie? Was it days or weeks since he had seen her?

'Millie, where are you?' he called out.

'I'm right here, James,' came the familiar voice. He could see her now, in the mirror. 'Look,' she said. He saw beneath her naked breast a beating heart, all bright colours.

Millie faded. His reflection was all that remained. Staring hard into the mirror, his face sometimes assumed a sinister aspect. With no discernible change in expression, it suddenly filled with menace and contempt. He would cover his eyes and shake his head to break the spell. But this time he was slow to react. The malice surged from the mirror and poured through his helpless eyes. It breached the wall of the retina and hurtled down the visual pathways deep into his brain. He felt it. He watched it all the way.

Too late. He was gone. Now there was no face in the mirror, or rather there were remnants of a face. There was nerve and muscle, bone and cartilage, lidless eyes and skinless lips. Abstract shapes. Light and shade.

He had searched in vain for a scintillating rock pool. The ones he found were bleak puddles compared to the picture in the encyclopaedia. Now James stood naked at the edge of the sea, the bones of his skeletal toes submerged in damp sand, his skeletal hand outstretched against white waves and grey cloud, salt spray fresh in his transparent nostrils. The chill of the water lapping at his ankles shot neuronal needles to his testicles, half pleasure, half pain. He spread his fingers for inspection – white bone, skin, muscle, veins – then bunched them into fists.

'I am James Moon, the Visible Man!' he said in a half-shout that was swept away by the ocean wind the moment the breath left his transparent lips. 'This is what I am!'

His thoughts leapt and curled within the folds of his brain. Sharp as lasers, they were casting shadows across the sky.

TWO

The Spark in the Stone

I Think Therefore I Am Dead

I am sitting alone in the small seminar room on the tenth floor. This is known as 'Harry's room'. I am at the head of a long oak table, working at a laptop computer. The door is at my back and the single window at the other end of the room sheds a thin, early evening light. There are glass-fronted oak cabinets along the walls, left and right. On the shelves are rows of display jars containing specimens of human brain, each suspended in a liquid the colour of watery piss. This is Harry's collection. The specimens are arranged according to pathology: tumours, cerebrovascular disease, degenerative disorders, and so on. There are whole brains, half brains, and parts of brain, sliced and segmented. Close to my right shoulder, there swims a cerebellum.

The room is ineffably still. Among the relics of natural disease and degeneration sit three victims of unnatural violence. Their stories intertwine. The first brain was caught with the second brain's wife and was dispatched with a pistol shot to the back of the head. After putting an end to its wife, the second brain finally dispatched itself. The woman's brain, third in line, completes the set. Hers is perfectly intact. She got it in the heart, according to Harry. I once told him I thought she might have been better placed between the other two, to keep the rivals apart.

'Even in death,' I said, 'you can sense their contempt for one another.'

It didn't seem to worry him. And, anyway, I wondered, what was she doing here? Her brain didn't illustrate a pathology of any kind. Harry's response was that she exemplified the normal, intact brain. He wouldn't concede that in displaying the specimens in this way he had also created a tableau, showcasing the fickle heart as much as the fragile brain. All the same, it was a tale he seemed fond of telling.

The material substance of the brain was bread and butter to Harry, a neuropathologist, but not to me. I remember the ambivalence I felt when I first held a human brain in the palm of my hand, the fascination but also the distaste. I was surprised, and moved, by how heavy it was. Perhaps a part of me had expected it to be weightless, like a mental image or a train of thought. I was eager to confirm for myself that the internal structures matched the familiar textbook pictures but, somehow, felt disinclined to start cutting. I imagined the worlds it had created: sky, clouds, people, pleasure, and pain. Everything. It's all in there.

'I could be bounded in a nutshell and count myself a king of infinite space,' said Hamlet, 'were it not that I have bad dreams.' The infinite space was within the shell of his head. And so, inescapably, were the dreams. But looking around now at these dead still, grey-beige objects it is hard to see them as erstwhile progenitors of infinite space. They each represent the opposite: a singularity. A point at which the universe has collapsed. I love the stillness of this place and the hum of the void – the sense of worlds dissolved and dissipated passions. It fills me with a sense of being. I am not yet pickled meat.

The light is fading and the pale amber sky at the horizon almost matches the colour of the liquid in the jars. There is a single, bright star.

My area of supposed expertise, neuropsychology, is the subject about which I feel the most profound ignorance. I am ignorant of many things. For instance, I know nothing of the Russian language. Quantum physics is beyond me. Keynesian economics? The workings of the internal combustion engine? Irish political history? My knowledge consists of vague notions poorly understood, loosely grasped general principles, and collections of disjointed facts. But I could take lessons in Russian and mug up on Irish history and the other things. I will never master the mathematics required for a proper understanding of quantum mechanics, but I can appreciate something of the flavour of the subject from the popular writings of experts in the field, and take comfort from the fact that, fundamentally, it seems to be beyond them, too. But when it comes to understanding the relationship between the brain and the conscious mind, my ignorance is deep and there is nowhere to turn.

An ocean of incomprehension heaves beneath the textbook-confident surface of plain facts and technicalities that I present to my colleagues and patients. I have a clear picture of the material components of the brain and am prepared to ad lib at length about features of its functional architecture – the interlocking systems and subsystems of perception, memory, and action. But quite how our brains create that private sense of self-awareness we all float around in is a mystery. I have no idea how the trick is achieved.

Wouldn't it be absurd for an airline pilot to deny knowledge of the principles of flight, or for a physician to claim ignorance of the basics of human physiology and anatomy? Yet I, a neuro-psychologist, can give no satisfactory account of how the brain generates conscious awareness. Worse still, I find myself edging towards a doubt that it means anything at all to say that the brain generates consciousness.

Hardly anyone visits Harry's room these days. There are small committee meetings once a month and an occasional journal club. Otherwise, it is used as I am using it now, as a quiet space for catching up with discharge letters and clinical reports. It was never used much for seminars and Harry, of course, doesn't come here any more.

I've been trying to finish a report. The patient, Jeanie, has a dementing illness and seems to be rapidly fading away. She is only fifty-three. It's not Alzheimer's, I'm pretty sure of that, but we have yet to come up with a firm diagnosis. I saw her this morning in a side room. She had been sharing a bay with five other beds on the main neurology ward until a couple of nights ago when she became agitated and began to develop delusional ideas about the other patients. A nurse found her at three in the morning packing a case and preparing to leave.

'I don't want to cause trouble,' Jeanie had said in a whisper, 'but I'm not like the rest of them. I shouldn't be here. They're all lesbians.'

This morning she was cheerful. An orderly had just brought her a cup of tea. Her daughter, Lisa, was feeding her baby on the other side of the bed. Lisa visited every day. We sat and chatted with sunshine streaming through the uncurtained window. Jeanie was happy, though increasingly preoccupied with thoughts of death.

'I've often wondered,' she said, 'what happens, medically speaking, when you die.' She wanted to know the procedure when a patient died on the ward. How could the doctors be sure someone was dead? Where did the body go? Who took it?

'We have work to do,' I said, 'shall we press on?'

First, I checked her orientation for time, place, and person. Fine. She knew who I was, where we were, the day of the week, and the month. She was quick to supply autobiographical information and seemed fully aware of her current circumstances. Next, I began to probe different aspects of mental function with some standard bedside tests. One of these was a verbal fluency task in which she had to generate words with a designated initial letter. The first letter was 'F'.

'Fire, flag, funeral,' she said. 'Will that do?'

'Tell me some more – as many as you can,' I urged her, but the allotted sixty seconds ran dry with nothing more to show. She managed just one word for 'A' and another three for 'S'. From letter fluency, we moved on to categories.

'Let's see how many different kinds of four-legged animal you can think of,' I said, and Jeanie pinched the bridge of her nose. Half a minute went by with no response. The baby, now asleep in her carrycot, began to stir, then settled. I reminded Jeanie of her task.

'You know,' she told me, 'I seem to be having problems with this one. Four-legged animals? For some reason I can only think of three-legged animals.' I noticed the trace of a smile on Lisa's lips, but her eyes were as dull as lead.

I realize that it might seem mad to question the role of the brain in consciousness. There can be no doubt that brains and self-awareness are in close alignment. My brain and I are never far apart, and I accept that I am sitting here, in Harry's room, with my living brain, conscious and self-aware, whereas those lifeless specimens in the oak cabinets are not. I am thinking thoughts, listening, and looking. I can hear occasional sounds of traffic from the street far below and, unexpectedly, faint ripples of harpsichord music from somewhere along the corridor. The taste of coffee is still in my mouth and I feel the contact between elbow and table, knuckles and chin, as I lean forward to read the text on the computer screen.

With conscious deliberation I have been stringing words together on the screen in front of me through the play of fingers on keyboard, intermittently catching and turning over unsolicited, idle thoughts and images. (At one point, I find myself humming a Bob Marley tune. It drifts in from nowhere.) And there, through the window, I see a star, a hundred million miles away, but simultaneously also in my head. Its image enters my eye and flow-charts through the visual systems of my brain, finds a link with memory and language and, from outer space, gains a name and a location in semantic space: 'Venus'.

So, does conscious awareness have a physical location: mine, here and now, in Harry's room, precisely somewhere between my ears? Self-evidently, it seems. But then, go into the skull;

visit the brain's interior workings and you will find that there is nothing much to see. Not a spark of colour or whisper of sound and no signs of intelligent life. As you wander through this silent land you can describe its geography adequately enough in the third person, but, quite obviously, not the first.

From this vantage point it seems self-evidently true that consciousness does *not* have a particular location. It is no more to be found in the hills and dales of the frontal lobes or on the slopes of the Rolandic fissure than in the chair you are sitting on. The more you search the terrain, the closer your analysis of substance and structure, the faster the will-o'-the-wisp recedes. We are embodied, but nowhere traceable within the physical structures of the body. I don't believe in immaterial mind stuff or souls detachable from bodies, and I'm not saying that the brain isn't necessary for consciousness. Whether it is *sufficient* is another matter.

Jeanie grew tired of my tests. She was losing concentration. In the middle of some mental arithmetic, she slowed to a stop and I let her sit and stare for a while. Lisa was sitting back in her chair, head resting against the wall, eyes closed. The baby was fast asleep. Hospitals are never quiet, but you find pockets of resignation and weariness where time itself seems becalmed. The sounds of the outside world are distant and abstract. We each withdrew into our private worlds. Jeanie, I allowed myself to imagine, was roaming some high plateau of bewilderment in pursuit of three-legged animals; the baby was drifting contentedly on a pond of mother's milk. But I did not presume to imagine what Lisa was thinking.

Consciousness is a puzzle. From one perspective it seems that it must have a physical location (people's pains and pleasures go where people go), yet, from another, the same suggestion seems faintly absurd. Once inside the head it becomes clear that consciousness is not a 'thing' to be located. And even if we think of it as a 'function' or a 'process' rather than a 'thing', what sense does it make to say that the crucial elements reside in this or that region of the brain? Nor does consciousness depend in some mysterious way on the integrated functioning of the whole brain. I have seen many patients who, as a result of surgery, injury, or disease have had much less than whole brains and they seemed perfectly conscious as far as I could tell. I'm sure they'd tell you they were.

For Wittgenstein, philosophy was not so much about finding solutions to puzzles as about correcting fundamental misunderstandings. The philosopher's treatment of a question, he said, is like the treatment of an illness. Our minds are knotted with misconceptions about the world and the job of philosophy is to unravel the knot, or, as he said, to show the fly the way out of the fly-bottle.

There was a time – before the brash intrusion of cognitive science – when the 'mind-body problem' lived quietly in the cloisters of academic philosophy, no trouble to anyone. These days, the redefined field of 'consciousness studies' is a garden of delights, swarming with philosophers and scientists of every stripe. Debate is lively, sometimes strident, and with the neuroscientists shouting loudest of all above their noisy brain scanners, most do not notice the fly buzzing frantically to escape the fly-bottle. They are engrossed. How *does* the mental arise from the material? How *can* subjective experience be reconciled with that soggy mass occupying the skull? They are full of confidence, too. Most of them expect a solution. The chimera of consciousness rises like a vapour and entices them to believe that it really is just a matter of time before a way is found of accounting for subjective, first-person phenomena in objective, third-person terms. Despite the prodigious amount of intellectual energy that has been driven into this enterprise in recent years, philosophical and scientific, it seems to me that the fly is still stuck in the bottle.

Eventually, Jeanie said, 'Am I dead?'

I didn't respond immediately. Lisa's eyes remained closed and I let the silence flow. Jeanie smiled. Her face was lit with a benign perplexity.

'I'm just wondering,' she said. 'Have I died?' There was a smear of toothpaste around the corner of her mouth. She didn't seem to notice the droplets of tea spilling on to her dressing-gown. But there was a glint in her eye. She was developing her theme. 'In the middle of the night I was convinced,' she said. 'I thought they would come to take me away. No, I wasn't afraid. I waited to see what would happen. And then someone did come. It was a tall man. He just watched, and I tried to say something, but my lips wouldn't move. Then the tall man left. He didn't say a word either.'

Lisa spoke. 'We've been through this before, Mum. You get confused sometimes. You're not going to die. Not for a long time.'

Given the uncertainty of her mother's diagnosis this was, of course, not necessarily the case. Jeanie gave no indication that she was listening.

'I can't say for sure that I am dead,' she continued, 'but things are not the same. I don't feel real. It seems to me I might be dead.' Her expression dimmed. 'How would I know if I was dead?'

Jeanie was well oriented for time, place, and person. She knew the day and the month, the name of the town we were in and the hospital, and she was clear about her name, age, and address. As for being dead or alive, she was all at sea. I wrote on my notepad: *Cotard's?*

I once saw an old woman who was profoundly depressed.

'Bury me,' she said. 'You might as well, I've been dead for some time.'

She believed her insides had rotted away. I tried to reason with her, but it was useless. 'Look,' I said, 'you're here talking to me. How can you be dead?'

'Just words,' she replied. A world of shadows flickered around her, human figures came and went, the curtains billowed, nights fell, days broke. But she felt no connection with any of this. Time hollowed her carcass and words fell dead at her feet. Just words. That was the first time I'd come across Cotard's syndrome, which is usually associated with severe depression, but is sometimes seen in cases of neurological disease. The person sinks into a nihilistic delusional state, often, as in this case, to the extent that they believe they no longer exist.

The condition takes its name from the French psychiatrist Jules Cotard who, in 1882, published a series of case studies of people suffering what he referred to as *le délire de negation*. The clinical presentation differed somewhat from patient to patient, but delusions of self-negation were common. These ranged from the belief that parts of the body were missing or had putrefied, to the complete denial of bodily existence. The expressed belief that one is dead is not a defining feature of the syndrome. In fact, of the eight 'pure' cases reported by Cotard (excluding a further three with concomitant persecutory delusions or other debilitating illness) only one embraced death as an explanation of her condition. Others slipped into non-existence, or skirted the abyss, somehow defying the conventional understanding that ceasing to exist must be tantamount to death. There were even some patients locked in the paradoxical state of denying their bodily existence yet at the same time believing themselves to be immortal.

What drives such strange delusions? Depression is usually a factor, but is not always present. Jeanie, peering quizzically into the void, is not depressed. Her case, at least, calls for a neurobiological explanation. One possibility is that the experiences arise from a disturbance of brain mechanisms which ordinarily bind sensation and thought to the neural systems underlying emotion. This ancient duty is performed by the limbic system, deep inside the cerebral hemispheres. A prime function of this system, an evolutionary *raison d'être*, is to create states of readiness for action. It does this through the implementation of so-called 'affect programs'.

If your sensory systems inform you that there is a crazed-looking man fast approaching with an axe, your body will enlist the affect program identified with fear. Before you have time even to experience terror, before the eye-bulging, voltage surge of awareness, various physiological systems will have reconfigured themselves in preparation for a response. You will turn and run. The thought 'I am terrified' will follow hot on your heels, though, most likely, will have entered the past tense by the time it catches up. 'I was terrified,' you will later recall.

But what is this 'I' that claims the terror, and what is the 'you' that reflects upon the experience? It is not a single thing, or a thing at all. It is, in its most primitive form, a principle of biological organization. The affect programs, so this story goes, not only guide adaptive interaction with the external world but, as a by-product of this process, they also form the biological point of origin of the self. By imbuing perceptions, thoughts, and actions with an emotional hue (however pale) they give cohesion to experience.

Feelings are generated which form the basis of our sense of identity, creating the conditions for ownership of thoughts and for agency in the control of actions. These perceptions, thoughts, wishes, beliefs, utterances, and actions are *mine*. I feel it. Their common cause is centred upon my needs and motivations, made manifest through the affect programs of my limbic brain. I *feel* I think, therefore I am. Note that this is merely a functional description of the biological roots of the self. Don't ask where the *feeling* of the feeling comes from; or the feeling of the feeling of the feeling. Such questions tighten the knot.

Beyond this unelaborated, biological core there are, of course, dimensions of the self with a past and a future as well as a raw present: in narrative terms, the autobiographical self. In Cotard's syndrome, however, the core has dissolved. Cognition is decoupled from feeling and, consequently, thoughts and actions have no fixed moorings. There is no 'I' left to claim ownership. It disintegrates; the fragments drift apart. One patient believed she had become little more than fresh air: 'Just a voice, and if that goes, I won't be anything.' If the voice went she would be lost and wouldn't know where she had gone, she said.

Jeanie became fascinated with her teacup.

'Look at this,' she said. 'Is it real? How can I tell? It doesn't look real.' She contemplated the object as if it had just materialized out of thin air, then her gaze turned to me. 'And what about you?' she said. 'Are you real?'

I had stopped taking notes and sat, hands clasped over my head, pondering the innocent question. 'Believe me,' I said, 'I'm real and so are you. Take my word for it.'

'I think I can trust you,' she said, but she wasn't sure.

Some philosophers (dismissed by others as 'Mysterians') argue that the 'problem of consciousness' exceeds human mental capacity in the way that differential calculus or the concept of democracy are beyond the intellectual scope of a rabbit or a pigeon. I find this view curiously comforting, but then I'm more of a clinician than a scientist. In my trade, unlike science, incorrigible optimism can be counter-productive. Some problems have no solution. But if there is a way to untie this knot of knots perhaps the first move is to acknowledge that we are not only physically *embodied*, but also *embedded* in the world about us. The mind may be local to the body and the brain, but it is also, in different ways, distributed beyond biological boundaries.

The notion of 'the extended mind' has been gaining currency in cognitive science, but similar ideas were developed more than fifty years ago by the Russian neuropsychologist Alexander Luria. For Luria, psychological phenomena were part of the natural world and so subject to the laws of nature, but he also recognized that the structure of the mind has social dimensions.

He thought that scientific psychology should be aligned with the biological sciences, but believed that one could never fully understand the relationship between the brain and the mind by treating the brain as a closed biological system. The working brain has to be understood not only as part of a larger biological system (the rest of the body), but also as a component of the wider social system. What we refer to as the 'self' is a product of biological and social forces arising from the *interaction* of individual, isolated, brains. There is no spark in a single stone but, struck together, two stones can start a blaze.

The challenge for neuroscience will be to fit the brain (a biological object) and the self (a social construct) within a common framework of understanding. The brain sciences may have to open up to a 'social paradigm'. Far from being the Holy Grail of neuroscience, the search for consciousness within the circuitry of an individual brain can lead only to fool's gold. Santiago Ramón y Cajal (joint winner of the 1906 Nobel Prize for his work on the structure of the neuron and one of the founding fathers of modern neuroscience) once said: 'As long as our brain is a mystery, the universe, the reflection of the structure of the brain, will also be a mystery.' We and the world are tightly intertwined. Though we may not have a special place in the universe, the universe, as far as we can ever understand it, has a special place in us.

'I think I can trust you. I think. I think . . .' Jeanie's words were struggling for life. Her gaze drifted over the pale-blue paint on the wall. 'I think I can.' Moving with a mother's grace, Lisa lifted the sleeping baby from the cot and placed the bundle of blankets and pink flesh in Jeanie's arms. Jeanie kissed her granddaughter and began to weep. It was time for me to go.

'Mum's more herself after a good cry,' said Lisa.

'That makes sense,' I told her.

Jules Cotard died at the age of forty-nine. He succumbed to diphtheria after nursing one of his children to recovery. I recall this fossilized fact of biography as I stack my case notes. The glow of the computer screen is now brighter than the sky and, when the machine shuts down, Harry's room is almost dark. Paradoxically, as the gloom descends, the jars along the walls gain a kind of luminescence, as if they have absorbed some of the receding light. My report is finished. The laptop lid closes with a satisfying click and I go across to take a closer look at one of the brain specimens. I lean close to read the printed label: *Subarachnoid haemorrhage.*

'How's it going, Harry?' I say.

~

How's it going, Paul?

Me?

There's no one else.

I was lost in thought.

What were you thinking about?

Nothing much.

The immensity of the universe, the mystery of consciousness, and the finality of death, no doubt.

Yes.

Fetch me a Gauloise!

It would be good to cause a stink.

I'd love a cigarette.

It must be torture.

Tell me more about the woman who thought she was dead.

I've nothing to add.

There's plenty more you could say.

But I'm not going to.

Why not?

Let the story stand. It's truer to life. I don't always know the final outcome – and that applies to Jeanie.

The diagnosis, at least. The prognosis.

Hashimoto's disease. Uncertain.

So there was hope?

It's an inflammation of the brain.

~

Vodka and Saliva

This afternoon I drove to the beach. There were no takers so I went alone, or rather, it was me and the dog. A dog is company if you don't think about it too hard, which mostly I don't.

It had blown up chilly by the time we got there, but I stripped to my shorts and went in, cautiously at first. The water was aggressively cold. The only way to proceed was not to think but to act, so I instructed my body to trot forward and dive into the next wave. Dutifully, it did, although I watched the approaching wave with trepidation. Under the water there is actually a dulling of sensation, as if consciousness itself is momentarily submerged in the thrum of the ocean, then it returns with full force. I surfaced and rolled on to my back, gasping with the cold, arms and legs driving the water, intensely aware of every startled neuron. I was enveloped by sea and sky, but now felt detached from both. The dog paddled beside me showing no signs of discomfort.

Descartes believed that dogs, indeed all animals, are unconscious automata. An animal screaming in pain is like the chiming of a clock. My faithful friend is a machine. Its fidelity is merely reflexive. It doesn't feel the cold. Intriguing, then, to learn that the great man himself kept a pet dog, Mister Scratch, of which he was very fond.

'I know that I exist,' said Descartes, 'the question is, what is this "I" that I know?' He was quite sure the 'I' that he knew was not his body. 'I am not this assemblage of limbs,' he said, but of course he knew he was. Just as, at one level, he must have believed that Mister Scratch had some degree of conscious awareness. He was far too clever to feel affection for an automaton, surely.

I might not be as clever as Descartes, but I trust my intuitions, and it seems to me that my body is an important part of the 'I' that I know. It is the physical apparatus over which I have direct control, the thing I urge to dive into icy waters, the thing that goes to work and sees patients and gives lectures. I never leave home without it.

My body has certain boundaries (roughly defined by my skin), which give it a characteristic shape; and as I steer it from one place to another my thoughts and experiences go with it. If you are having a bad time for some reason and I say 'My thoughts are with you,' don't believe me. My thoughts are very much with *me*. Always. Believing that thoughts are displaced from your body or that other people's thoughts can be inserted into your head, is a sign of mental illness.

My body is, without doubt, a part of what I think of as my 'self'. It's the part of my self that can be weighed and measured; it casts shadows, and it has properties in common with other physical objects like trees and filing cabinets, cars, and planets.

'Body is a portion of the soul discern'd by the five senses,' said William Blake.

I have a strong sense that I am located *in* my body. I drive my car to the beach and I drive my body into the cold shock of the waves. On the way to the beach I see hedgerows and trees flash by through the windscreen of the car and, trotting into the water, I see the waves and the sky as if from behind the windscreen of my eyes. I feel located in my body and I identify with it in other ways, too. For example, if I see it among other bodies pictured in a photograph I might say something like 'That's me' or 'There I am.' I'd say something similar even about an old photograph showing me as a baby, despite the fact that the body bears no resemblance to the one I currently have.

If someone passes my body in the street they might, if they recognize it, offer a greeting, using my name. A name is another way we have of thinking about our selves – a label to identify our bodies and mark their actions. 'That's Paul over there, running into the sea.' One can change one's name, but not one's body.

So, I *feel* I occupy my body (there is no stronger intuition) and, with that, comes a sense of *ownership* and *agency*. It's *my* body and *I* control it. I make it *do things*. My body also contributes to my sense of continuity – the feeling that I am the same person from one day to the next. When I look in the mirror each day I expect to see the same thing, more or less. I'd be surprised if one day I looked in the mirror and saw Nelson Mandela or a woman or a giant moth. I'd be rattled.

Identifying the self with the body seems reasonable enough, but there are some problems. For example, the boundaries of the body are not so easy to define. How much a part of us are our hair or our fingernails? What about bodily fluids? What about food? I pick a strawberry from a basket, I swallow it and it becomes incorporated into my body. At what point does it become a part of my body and so a part of me?

As a student I had tutorials with the famous psychiatrist Anthony Storr. He was a relaxed teacher, very charming, and I'm sure I learned something about psychotherapy. But all I can recall is one of his thought experiments.

He asked us to consider how often we swallow our own saliva. We do it all the time, of course, without thinking. Then he invited us to imagine that, instead of swallowing, we spat into a tumbler. How would we now feel about sipping from a tumbler full of our own spit? It's the same stuff, but *no thanks!* Not even with ice, lemon, and a large dash of vodka. What's the difference? A boundary has been crossed. As the philosopher Daniel Dennett puts it, once something is outside our bodies it becomes alien and suspicious, not quite part of us, something to be rejected. The spit in the tumbler has 'renounced its citizenship'. Boundaries and border controls are important.

Dennett also reminds us that the society of the human body has many interlopers – bacteria, viruses, microscopic mites – not all 'enemies within' or even tolerated parasites. Some, like the bacteria in our gut, are vital to survival. I identify with my body, but not with any of these bugs, or items of food passing through my digestive system or, indeed, with any particular *part* of my body on a larger scale – my knees, my knuckles, the blood coursing through my veins. I could lose an arm or a leg or a pint of blood and I would still be me. Perhaps it's the *idea* of having a body that really matters.

I may feel that I inhabit and control a body and that such feelings are fundamental to my sense of self, but there are many features of my body over which I have no direct control. I can't stop the ageing process. I can't stop it developing a tumour or a degenerative brain disease, if that's what the genes dictate. And there are millions of physiological processes going on inside me that I scarcely know about, let alone control.

Although I can claim a better than average knowledge of human biology, I have only a general notion of what my internal components are. Many intelligent people with a perfectly functional sense of self haven't a clue about what goes on inside them. It is largely irrelevant to the everyday business of being a person. Just as when you drive a car you don't really need to know how the engine works.

Even when you consider those things that we directly take charge of, the activities of the body through which we exercise our free will (voluntary movements of the limbs, fingers, head, vocal apparatus, etc.), even here, the degree of control is sometimes so poor that we achieve effects in the world quite opposite to those we intend. The practice of deception is a case in point. When people display expressions for emotions they are not feeling, or say things inconsistent with their actual state of mind or their true beliefs, there are often counter-signals that give them away. This applies whether we are lying or, for the best of reasons, simply trying to give a false impression to disguise the true state of affairs.

Paul Ekman, a pioneer in the study of emotional expressions, lists some of these tell-tale signs: 'a movement of the body, an inflection to the voice, a swallowing in the throat, a deep or shallow breath, long pauses between words, a slip of the tongue, a microfacial expression, a gestural slip . . .' Lies can be performed beautifully, says Ekman, but usually they are not. And then there are occasions when we behave with perfect control of our actions, but our behaviour is, at some level, not what we wish or intend. We act against our better judgement; we yield to temptation.

When I finished that last paragraph I got up and went to the lavatory. I have absolutely no idea how I did it. I became aware of an 'urge' to go, I stood and found myself walking to the bathroom where, magically, effortlessly, I hosed urine into the toilet bowl. Don't ask me how. I take it for granted that I can just 'think it and do it'. The co-ordinated neural, musculo-skeletal and urogenital activities involved in the enterprise of getting up and going to the lavatory are incredibly complex. I just made it happen. I have phenomenal control over neurobiological processes that no one in the world fully comprehends, and I don't even have to think about it.

It reminded me to make the point that even when we have excellent control over our voluntary actions, and at every level intend to perform them, we still don't understand precisely how an act of will gets translated into a complex sequence of biological activity (or vice versa).

So, we can see that the body is an important feature of the way we think about ourselves – it seems natural to believe that each of us owns a body and that we have control over it. But we can also see that it is difficult to identify the 'self' with the 'body' as a whole (because the boundaries are fuzzy), or with any particular part of the body. Furthermore, our control over our bodies, and our understanding of the processes involved, is variable. Perhaps, as I say, it's the *idea* of having a body that really matters.

One might think that the face has some special alignment with the self. No other object projects such an aura of vitality, and this vitality seems to come from within. Faces are points of convergence between people; where we seem to locate the essence of another person, and where we tend to locate *ourselves*: somewhere behind the eyes. In his novel *Immortality*, Milan Kundera writes: 'Without the faith that our face expresses our self, without that basic illusion, that arch-illusion, we cannot live or at least we cannot take life seriously.'

Imagine if someone you know were suddenly to undergo a radical transformation of his or her facial features. They still have a face, a regular one, but a *different* one. Is it possible to believe it's the same person? What if they now look just like someone else you know? Or what if they resemble you? Now imagine that person with no face at all. Can you even think of them as a *person*? What is it you are thinking about?

We treat the face as an emblem of the self. It generates potent illusions. Kundera might be right to say it would be hard to function as a human being without embracing the emblem and seeing the illusion. But it would be a mistake to identify faces with selves. The face is just another body part. People with horrendous facial disfigurements have no less a sense of self than people who have lost an arm or a leg. In some respects, perhaps, their sense of self is intensified.

The face is just a fleshy structure animated by muscles attached to the bony structures of the skull. It contains information about our identity (who we are), our sex and our age (which are important facets of the 'public self'), who and what we are in terms of the objective, social facts of the matter. You can think of these as 'static' features of the self in so far as they are relatively fixed and enduring.

Then, through changes in patterns of muscular activity ('expressions', 'gaze'), the face transmits signals about other, more dynamic, features such as our emotional state, our focus of interest, and our immediate intentions. These have a double aspect; part public, part private. You can use facial information to make inferences about my mental state and behavioural dispositions. To that extent the information is 'public' because it is there for anyone to see. But you can't know my thoughts and feelings directly. You can't experience them.

We see and hear and speak through the face, creating the impression that consciousness, 'the stuff of the self', is concentrated there, even though there are no grounds for believing it is really any more 'there' than in the right elbow or the small of the back. This is because there is no 'self stuff' to be located. There is nothing in, or behind, the face except for organic matter, and nothing to suggest that the biological material of the head, as organic matter, has a greater propensity for 'selfhood' than the material stuff of other regions of the body. There just *isn't*.

Travelling in thought from the position of participant-observer in the physical and social world 'through' the face and into the machinery that lies behind we are transported, like

Alice through the looking glass, to a very different world. We go from a bright place of persons, selves, and subjective experience, to a dark, silent, enclosed, world of physics, chemistry, and biology. It is a mysterious journey.

Body Art

There's someone here to see me. She's come to talk about her research project. She's looking for a PhD supervisor.

'Hi, I'm Kara,' she says, drifting in like scented smoke.

'I'm more of a neuro man,' I'd told her over the phone. 'I'm not sure I can help.' I tried to put her off. I said I knew something about body-image changes caused by brain damage, and self-mutilation in the mentally disturbed, but nothing about the cult of extreme body modification. She wouldn't be deterred, and here she is, opening a folder to show me some samples.

The thing that first catches my eye is a close-up colour photograph of a man with his tongue hanging out. Tongues, almost. It is split from the base, giving it a wicked, reptilian look. You can almost see it flicker. Kara has a glistening stud in the middle of her own tongue. It's difficult to ignore once you notice.

The tone of the Information Sheet is reassuring. It could be from a private hospital brochure. I learn that *The most popular method of tongue splitting is surgical.* Images of DIY enthusiasts with razor blades and scissors rapidly fade. Nothing of the sort.

The operation is quick and high-tech, *performed by an oral-maxiofacial* (sic) *surgeon using an argon laser.* The tongue is slit in a single sweep, the laser cauterizing as it cuts. Long-term side effects are played down. There may be minor changes in some speech sounds, it says, and, physiologically, the number of taste buds increases to cover the extra surface area. Elsewhere, a woman reveals it took about three weeks before she could eat comfortably and control *both tongues.* No claims are made for the gastronomic or sexual advantages of the split tongue, but I begin to wonder.

There are many other images. Some are relatively mundane (tattooed penises, nipple piercings, scarification, branding) and some bizarre, like transdermal implantation. Kara shows me pictures of men with objects inserted into the forehead or scalp. They look like *Star Wars* characters. I cast my eye over a report on *non-psychotic self-cannibalism (autophagy)*, and another on *apotemnophilia*, which, I learn, refers to a craving for amputation, sometimes satisfied through the services of qualified surgeons.

Body art has filtered into the mainstream. Almost everybody has a tattoo or a piercing these days. The same goes for body modification: breast implants, nose jobs, liposuction, anorectic dieting, body-building. Kara condemns it all as a hopeless striving for unobtainable ideals of conventional beauty and eternal youth (the women in *Vogue*, the men on the cover of *Men's Health*), culturally sanctioned and commercially driven. She's right, of course. Extreme body modification, however, is the antithesis. It's about redefining the aesthetic, even the boundaries, of the body.

'And what about circumcision?' she says as an afterthought. Kara, by the way, is distractingly beautiful.

Searching for common ground, we skim across body-image distortions in neurological disorders such as epilepsy and stroke. We discuss hysterical paralysis, phantom limbs, and transsexualism. I tell her about anosognosia, which means 'lack of knowledge of illness'. People with severe neurological disabilities — quadriplegia, say — sometimes show a complete lack of awareness of their condition. (I remember once chatting with a man who was paralysed from the neck down. He was telling me about his plans to go rock climbing at the weekend.) I agree that 'body image' is a fascinating area for research, but can't immediately see a point of connection between Kara's interests and my own. I tell her I'll think about it.

At home, I stand naked in front of the bathroom mirror. Not exactly *Men's Health*, I think. What might a little body art do for me? I tell my wife I'm thinking of having my penis tattooed.

'What do you have in mind?'

'*Wolverhampton Wanderers.*'

She looks at me. 'Or maybe just *Wolves.*'

The Story of Einstein's Brain

Einstein: shock-haired and sockless genius, avuncular symbol of pure intellect, head in a whirlwind of equations and spiralling galaxies, cultural icon. Logo: $E=mc^2$. Scientist, sage, humanitarian, ambiguous pacifist, lousy husband, negligent father, and now, what's left of him, fragments of brain in a jar.

Not long after Einstein's death in 1955, Roland Barthes called his brain a mythical object, a paradoxical conflation of man, magic, and machine. Nearly fifty years on, the myth remains potent. To look at any brain is to confront a deep mystery. You fall into the frame of an impossible picture, an Escher stairway, ascending and descending at the same time. The brain can't be the theatre of consciousness – it's a solid object – and yet it must be because you are contemplating the scene on the floodlit stage in your own head. But, looking at photographs of Einstein's brain – snapped in the interlude between extraction from the cranium and decimation at the hands of Princeton Hospital's duty pathologist, Thomas Harvey – you feel the pull of myth as well as mystery. It is difficult not to see the object as a sacred relic. This is the thing that bent the universe and humbled time.

There was rumour and speculation about the brain from the start: it was huge and strange, and then again it was mysteriously tiny, the size of a walnut. In fact, it looked ordinary and weighed 2.7 pounds. About average. It was removed within seven hours of death, weighed fresh, then fixed in formalin. After it had been photographed from all sides, and measured with callipers, the cerebral hemispheres were separated and diced into 240 blocks. Then the brain disappeared. It followed Harvey into obscurity.

Soon after the autopsy, Harvey had announced that Einstein's brain would be used for scientific research and there was a tussle for possession between Princeton and New York's Montefiore Medical Center. They both lost out. Harvey simply took the pieces home with him and stored them in cookie jars. Never mind his lack of qualifications for the job (he was a clinical pathologist, not a neuroscientist), he would be the one to unlock the secret of Einstein's brain.

The pathologist was accused of a smash-and-grab exercise and, though Harvey always maintained he had acted on the authority of Einstein's executor, Otto Nathan, not many believed him. Nathan called him a thief and a liar and Harvey eventually left Princeton under a cloud, allegedly fired for not relinquishing the brain. He vanished from the scene – but he hung on to his cookie jars. In 1978 Harvey was tracked down by Steven Levy, a journalist working for the *New Jersey Monthly*. Levy found him living in Wichita, Kansas. The remnants of Einstein's brain were in a box marked *Costa Cider*.

Two decades later another journalist turned up: Michael Paterniti. The Keeper of the Brain was then well into his eighties and living in a basement back in Princeton. Together they set off for California in a rented Buick Skylark with Einstein's brain stashed in the trunk, floating in a Tupperware container. Paterniti wanted to explore rumours that it might be cloned, or sold to Michael Jackson for millions of dollars, but Harvey wasn't saying very much.

They ended up at the home of Einstein's granddaughter, Evelyn, who seemed less in awe of the relic than anyone else. Paterniti's own reactions were complex. 'I never thought that, holding Einstein's brain, I'd somehow imagine eating it,' he said at one point. Then, at a seedy motel on the way home, he slept with it: 'I go to bed. I put Einstein's brain on one pillow and rest my own head on the other next to it, six inches away.'

What about the science? Was Einstein's brain in any way extraordinary? Despite Harvey's pledge, no study was conducted for three decades after the contentious autopsy. By now he had begun to mail bits of brain to prominent neuroscientists, people better placed than he to examine the material.

Four sugar cube-sized pieces arrived at Marian Diamond's Berkeley office in a mayonnaise jar. She examined the cellular structure of the specimens microscopically, finding an unusually high ratio of glial cells to neurons in the inferior parietal lobe, an area known to be associated with mathematical and spatial reasoning. Neurons are the basic functional units of the brain and the glia provide the metabolic and structural support required for them to do their work.

As for overall anatomy, the first study appeared in 1999.

Sandra Witelson of McMaster University, Ontario, had received, unsolicited, a package of brain pieces which she and her colleagues set about weighing, measuring, and comparing with other brains. Again, the inferior parietal lobe stood out as unusual, being 15 per cent larger than normal; and the Sylvian fissure, which marks the temporal-parietal boundary, took an odd upward turn.

Such observations have been dismissed in some quarters as little more than primitive, bump-fondling phrenology. Einstein, they say, would have been appalled by the crudity of the science. I am not so sure. There is a picture of the Great Man undergoing EEG brainwave recording – his head an explosion of wild hair and electrode leads – while he is being asked to 'think of relativity'. He was clearly game for a laugh. And the bump at least has a plausible location given Einstein's mathematical prowess and what we already know about the organization of brain functions. It is something.

Some years ago my young son and I were in a shoe shop in Cambridge when in came Stephen Hawking in his motorized wheelchair. No present-day scientist matches Einstein's celebrity, but Hawking comes closest. Like Einstein he symbolizes pure intelligence. The shining mind in the shrivelled body has entered the popular imagination. I've yet to see his wasted shape on a T-shirt or his mug on a mug, but he has appeared in episodes of *Star Trek* (rubbing shoulders with Einstein and Newton) and *The Simpsons*. As Roland Barthes remarks, being turned into a cartoon is a sign that one has become a legend.

There was a woman helping him and they were looking at a rack of cheap trainers. Hawking didn't seem very interested, though it was hard to tell. My son went close up and stared and I expected him to say something indiscreet, but he lost interest. The trainers he was eyeing were a cut above Hawking's. He was not impressed, though I felt peculiarly touched. This was the man who visited black holes from his wheelchair and surfed event horizons. And he did so in Tru-Form trainers. Barthes would have liked that. It would have signified something.

And now Einstein's brain is back at Princeton Hospital. Actually, not as such. It has a new owner, one Elliot Krauss, pathologist. He keeps it in a jar somewhere secret.

~

They're perpetuating the myth.

Harry?

I didn't mean to startle you. But they are, don't you think?

The paradoxical conflation of man, magic, and machine?

Quite.

Neuroscience thrives on paradoxical conflations. Conflating mind and matter seems paradoxical to most people.

Yes. It's hard to equate mental life with the sludgy stuff of the brain. We should never lose sight of the fact that the brain is a dollop of mush. I should know.

Okay – man, magic, machine, and mush.

But trying to explain the genius of Einstein by
measuring his bumps with callipers! It's phrenology.

There's only so much you can do. He's been
dead half a century. His brain is all in bits.

I am thankful for small mercies.

~

Articles of Faith

neuropsychology: *noun [mass noun]* the study of the relationship between behaviour, emotion, and cognition on the one hand, and brain function on the other.

The New Oxford Dictionary of English

Articles of faith:
1. The brain is the organ of the mind.
2. The mind is modular.
3. The modularity of mind is reflected in the workings of the brain.

The brain is the organ of the mind. No one doubts that the brain is the root of all behaviour and experience. If you blow out the contents of a person's head – as schoolboys used to blow out the contents of birds' eggs – you are, likewise, left with an empty shell.

The mind is modular. Mental life is diverse and divisible. The mind is not a monolith. We distinguish the colour of an apple from its shape, weight, and texture as we lift it from the fruit bowl; and as we take a bite we separate the snapping, crunching sound from the taste of the juice. Then, looking back on the experience, we segregate raw sensation from the images we hold in memory.

Perception and memory are just two domains. The mind is a much broader confederation. There is also reason, emotion, language, motivation, and action. These facets of mind function independently, at least to some degree. It is possible to find malfunction in one domain alongside normal operation in others. An amnesiac appreciates all of the sensory dimensions of eating an apple, but has no recollection of the experience an hour later. Then again, someone with diminished senses but memory intact – a blind person, say – has no difficulty remembering. None of this offends common sense.

The modularity of mind is reflected in the workings of the brain. Mental functions are biologically compartmentalized. Different brain systems subserve different psychological functions. It follows that specified damage to the brain has predictable functional consequences.

* * *

mind: *noun 1* the element of a person that enables them to be aware of the world and their experiences, to think, and to feel; the faculty of consciousness and thought . . .

brain: *noun 1* an organ of soft nervous tissue contained in the skull of vertebrates, functioning as the coordinating centre of sensation and intellectual and nervous activity . . .

self: *noun (*pl. *selves)* a person's essential being that
distinguishes them from others, especially considered as
the object of introspection . . .

The New Oxford Dictionary of English

A foreigner visiting Oxford or Cambridge for the first time is
shown a number of colleges, libraries, playing fields, museums,
scientific departments and administrative offices. He then asks
'But where is the University? I have seen where the members of
the Colleges live, where the Registrar works, where the scientists experiment and the rest. But I have not yet seen the
University in which reside and work the members of your University.' It has then to be explained to him that the University is
not another collateral institution, some ulterior counterpart to
the colleges, laboratories and offices which he has seen. The
University is just the way in which all that he has already seen is
organized.

Gilbert Ryle, *The Concept of Mind*

The self has no location, however natural it seems for us to
believe otherwise. Ryle was reconfiguring the 'mind-body
problem', the ancient mystery: how do mental events arise from
physical substance? His suggestion was that, contrary to the
assumptions of many philosophers and psychologists, it was a
mistake to put mind and body on the same plane of analysis – a
'category mistake'. Just as the stranger could not find 'the
university' beyond the labs, offices, and playing fields, so we are
hard put to discover any trace of a conscious mind, or self, in the

brain. There is no ghost in the machine. Minds are the product of brains, and selves depend upon minds, but they require different forms of understanding.

I am using a personal computer to type these words. They appear on the screen by virtue of the word processing software, essentially a set of instructions installed in the computer. The operations of the software are realized through the hardware of the computer's electronic microcircuitry. Detailed knowledge of the hardware is of little help in understanding the software, and vice versa. Both hardware and software are irrelevant to the content of the text. I happen to be writing about minds, brains, and selves, but it could be anything – a guide to sea fishing, a suicide note or a Japanese haiku. Think of the brain as the hardware, the mind as the software, and the self as the text on the screen.

In fact, why not a haiku?

A true enigma:
The self looks inward and finds
Nothing but neurons.

No more haikus, I promise.

* * *

Like the symbol on a dollar bill, my eye floats above a pyramid. The four sides of the pyramid represent the person, the mind, the brain, and the world.

When I'm with a patient I'm aware, at different times, of each side of the pyramid. Mostly, my attention is drawn to the *person*. It is a person who has come to see me or who is visited by me on a hospital ward. It is a person who reports a symptom of some kind ('I'm having problems with my memory'; 'I can't concentrate on anything'; 'I break down in tears if someone says a kind word'). There is always an 'I'. And even when – especially when – the 'I' is deformed by injury or disease, when it is submerged or dispersed and has no voice, I strive to make it visible and coherent. This is as much for my benefit as theirs.

And then I glance across the plane of the *mind*. 'You're having problems with your memory?' I say. 'Tell me, in what way does your memory let you down?' I question and probe, seeking clues to the nature of the problem. I know that memory, and therefore memory disorder, takes many forms. I can use special tests to help define and quantify the disorder. It is also important to know whether other components of the patient's mental apparatus are showing signs of wear and tear.

The confederation of mental processes we call 'the mind' can break down in ways not always evident to the person whose mind it is. They might complain of memory impairment, but unknown to them there could be other problems: subtle changes in perception, say, or reasoning or emotion. I take note of their symptoms, but all the while I am looking for other signs.

Next, my gaze shifts to the third surface of the pyramid. This is when I consider signs and symptoms in relation to the workings of the *brain*. I might, for example, take certain failures of memory to indicate a particular form of brain disease. Or, conversely, I might use knowledge of a patient's brain disorder to guide my understanding of their psychological condition. My observations are set alongside other forms of evidence, including the physical investigations of neurologists and surgeons, and the images made available via brain-scanning machines.

The fourth side of the pyramid represents the *world*. Here, I am concerned with how the person, given their brain disorder and mental profile, can best adjust to the world around them. How will they get by? For the person concerned this is, of course, the only question that matters.

~

You sound like an expert.

I am.

So you didn't mean what you said before: that neuropsychology was what you felt most profoundly ignorant about. It was a rhetorical shimmy.

Not entirely. It's easy enough to clip definitions and basic assumptions together. Like I said, I could ad lib about the structure and functions of the brain – no problem. And, true, those Articles of Faith get me through the working day.

But?

There's something quirky at the philosophical centre of neuropsychology. An incompleteness. The Articles of Faith suggest an integrity that, I fear, doesn't actually apply.

No science is whole. Physics hasn't got a Theory of

Everything. Einstein died intellectually frustrated.
If any science arrived at a state of completion,
the scientists would lay down their tools. Job
done. And there are definitely strange things
at the philosophical heart of modern physics.
Why should you wrestle with doubt?

For all I know, physicists will one day formulate
their Theory of Everything. But I'm inclined
to think that our goal of describing mental life
in terms of brain activity is not entirely feasible.
I wonder if the enterprise is quixotic.

Why?

Because . . . Oh, I don't know. What was that
description of Don Quixote? A muddle-headed
fool with frequent lucid intervals? Get back to
me when I'm lucid.

But does it stop you doing your job?

No.

Then why play The Knight of the Doleful
Countenance?

Have you read Cervantes?

No.

Me neither.

Are you saying that the Articles of Faith are fine
as far as they go — but they don't go far enough?
Something is lacking?

My concern is that the Articles of Faith disregard
some important features of mental life, which,
if we are ever to achieve a coherent science of

mind and brain, will either have to be brought into the frame of neuroscience – or thrown out.

Such as?

The things that matter most to us: conscious experience and our sense of self.

Is neuropsychology concerned with such things?

Of course. I've made a living by virtue of the fact that the brain is a very flimsy construction. Its functions are easily warped by disease and injury. There's no shortage of tales to tell of fragile brains and shattered selves.

Indeed.

The question is, how best to tell them: as the science of the brain or the art of being human? The hidden contraptions of the illusionist or the illusion itself?

Surely, for a clinical practitioner, both perspectives are necessary. Sometimes you are talking about the brain, and sometimes about the person, the self, consciousness and all.

One has to be bilingual, switching from the language of neuroscience to the language of experience; from talk of 'brain systems' and 'pathology' to talk of 'hope', 'dread', 'pain', 'joy', 'love', 'loss', and all the other animals, fierce and tame, in the zoo of human consciousness.

Then you seem to have tied the package: brain and person complete. Where's the strange philosophical centre?

I have come to realize how deeply odd it is to assume that brains and selves converge.

You think they don't go together?

I think they don't go together in ways that contemporary neuroscience would recognize.

But they go together?

What do you take me for?

~

Right This Way, Smiles a Mermaid

Midtown Manhattan. The power was out. I stood at the window as veins of lightning crackled over the General Electric building a block away. Its giant, galvanic throes were startling. The thunder rolled through my gut as it rattled the windows and humbled the monumental architecture. And then, an ocean of rain. New York was Atlantis. I could see fishes and whales and mermaids. When the storm died the power was still out.

I lay on the bed, drifted into a slumber and woke to find a girl standing at the window, looking out as I had looked out upon the storm.

'Don't be afraid,' she said, still watching the rain, which was gentle now. 'Come with me.'

We left by a door I hadn't noticed before. I followed her down dim-lit corridors and vertiginous stairwells, then out into the swirling street where a car was waiting.

'Right this way,' she smiled.

We were somewhere on the Upper East Side. I recognized

her at once, even in the gloom: Collicula Brodmann, President of the Academy.

'Neuroscience is a broad church,' she said, 'but there is concern that you may be drifting towards Mysterianism.' Out the window a blue whale dipped majestically over Central Park.

'Oh?'

'The Disciplinary Council has received a complaint.'

I took the envelope and read the letter. It accused me of bringing my profession *and the broad church of neuroscience* into disrepute (the same phrase) on account of my *anti-scientific posture* and *espousal of Mysterian philosophy*.

'Does this constitute a charge of some kind?' I asked.

'No,' said Collicula, 'but there are procedures.'

'Am I to be charged?'

'That's a matter for the Investigatory Panel. They will first consider your response.'

Three solemn figures entered the room.

'*Now?*' I said.

'Yes, if you are willing.'

It was hard to tell who was speaking. The three were seated in shadow some distance away. Then the man on the left – Number 1 – drew a candle to him and started to read from a file. I recognized my own words: *My area of supposed expertise, neuropsychology, is the subject about which I feel the most profound ignorance.*

He looked up from the file. 'Ignorance?'

'Profound ignorance,' I confirmed.

'If I were a patient of yours, would I be comforted to hear your proclamations of ignorance?'

'Professionals should acknowledge their limitations,' I said.

The candle passed to Number 2, a woman. She began reading from her file. My words again: *I find myself edging towards a doubt that it means anything at all to say that the brain generates consciousness.*

There was a long pause which, perhaps, I was expected to fill.

'You are a professional neuropsychologist?'

'Yes.'

'Have you ever been certified insane?'

Number 3 was a man: *Far from being the Holy Grail of neuroscience, the search for consciousness within the circuitry of an individual brain can lead only to fool's gold.*

He was direct. 'You believe that the relationship between mind and matter is unfathomable. In other words, you are a Mysterian.'

'No. I wouldn't be so bold.'

'And yet you find comfort in Mysterianism.'

'I am a clinician. I have it ingrained in me that some problems have no solution and that there are times when it is wise to accept the fact. As Wittgenstein said, the philosopher's treatment of a question is like the treatment of an illness. But if the disease is incurable, then so be it. I'm comfortable with the idea of not having solutions to every problem. I guess there's also a part of me that likes mystery for mystery's sake. Omniscience would be insufferably tedious.'

'And as far as consciousness is concerned, the disease is incurable?'

'Could be. I don't know. I'm indifferent to the mind-body problem.' This was not true; or rather it was not the whole truth.

'But do you, or don't you, believe that neuroscience can find the solution?'

'I'm not sure neuroscience has even found the problem,' I said. 'The fly is still stuck in the bottle.'

Collicula offered me a glass of wine. 'What do you believe?' she asked.

'Nothing,' I replied. 'Sometimes I wonder "How does meat become mind?" and it seems absurd.'

'Indeed.'

'Then, other times, I see it as a pseudo-problem, a screen of confusion...'–I realized we were sitting at a table, eating dinner. I had food in my mouth. I chewed and swallowed before finishing the sentence – '... behind which there is an empty space.'

The food was good. The wine was good. Collicula, I now noticed, was naked. So was I and, before long, we were making love; she writhed warmly beneath me on the glassy floor. There was an aquarium below with sharks gliding and smaller fishes darting. *How do I know this isn't a dream?* I wondered.

Now I was standing, naked, before the three solemn figures. I seemed to be giving a presentation. I looked at Number 1 and said: 'Some people believe that the universe and everything in it, including human minds, is made of physical stuff.'

'Materialism,' he said.

I turned to Number 2: 'The opposite view is that reality is non-material; physical objects and events are manifestations of mental activity.'

'Idealism,' she said.

'They believe the physical world is a figment. The universe exists entirely on a mental or spiritual plane.'

'Hmm.'

'Some versions of idealism don't deny the physical world, but say we can't have direct knowledge of it. Objects and events are mental constructions because they come to life only in the arena of the mind.'

'The unobserved tree falling in the forest makes no sound?'

'Yes. And there are no green leaves on the branches or dappled sunlight on the forest floor. The creaking, crashing sounds of a falling tree, the image of leaves and greenness, and notions of sunlight and dappling are all constructions of the mind.'

Number 3 was about to speak, but I cut him short. I had to press on: 'The third option is dualism. Dualists believe that the world is composed of both physical and mental stuff.'

'Dualism is dead in the water,' said Number 1. 'Modern science has no place for dualism.'

'But intuitively it feels right,' I added. 'Even to materialists and idealists who reject the idea intellectually. Even to you, perhaps.' He did not dissent. 'Every normal person believes they have a body and most tend to think there is more to them than that. They feel they have mental qualities distinct from their flesh-and-blood physical apparatus. Many people – probably most – believe they have souls that will survive the death of the body.'

'I hope you are not going to defend a belief in souls and spirits,' said Number 2.

'Certainly not.'

Number 1 returned to his earlier question: 'Are you a Mysterian?'

'No.'

'Then you accept that science will solve the puzzle of consciousness; it is merely a matter of time?'

'And research funding,' interjected Number 3, to the amusement of the others.

'No,' I said.

Number 2 told me I was confused. Number 1 wondered whether I had misheard his questions. I told him I'd heard him perfectly well but that, with respect, his questions were simplistic.

I knew what I wanted to say. I wanted to say something about the problem of consciousness being built into science itself, but I hadn't thought it through and, suddenly aware of my nakedness, was losing the thread of my argument.

'Let's go back to Descartes,' I said, as much to myself as to the others. I was looking for a way through.

'Must we?' This was Number 2.

'Yes we must,' I said. 'We must.'

I strode over to where the three were seated, leaned forward, and rested my elbows on the table directly in front of Number 2. 'He has a lot to answer for.' She smiled, rather sweetly I thought.

I reminded them that the mind-body problem, the beast we grapple with today, is a legacy of the dualist ideas formulated by René Descartes in the seventeenth century. He was not the first philosopher to distinguish between mind and body, but he crystallized that distinction and so set the terms of all subsequent debate about their relationship. In the process he released a pack of troublesome dichotomies into the Western way of thinking: mind versus matter; subjective versus objective; observer versus observed.

'You can't blame it all on Descartes,' said Number 3.

'Of course not,' I said. 'Dualistic thinking runs through every major religion – they all promote the fallacious idea that body and soul are separate entities.'

Number 1 pointed out, correctly, that the idea almost certainly predates organized religion by many thousands of years. He said it went back to the dawn of human history. I agreed. In fact, I believed it was part of our biological make-up. I said that, quite probably, we were innately predisposed to think in terms of the separation of minds and bodies. The idea was built into the hardware of the human central nervous system.

Evolution has endowed us with brains that are naturally inclined to certain ways of thinking about people, especially when it comes to interpreting their mental states. It was a consequence of living in complex social groups, and a by-product of the evolution of language. We continually, and effortlessly, picture each other's thoughts and intentions. We form assessments of what people 'have in mind' – presupposing that there are such things as minds. We are all mind-readers. And the same mental machinery enables us to form an idea of ourselves as unified and continuous beings – to make sense of what is going on with regard to our own mental states. People with impoverished mind-reading skills (such as autistic people), or with rich but unreliable interpretations of their own and others' mental activities (like schizophrenics) are severely disadvantaged.

What Descartes did, in effect, was to take this primordial intuition – the separateness of body and mind – and build a system of philosophy around it. And the ideas he formulated have become ingrained in our way of thinking. His division of mind and matter, and the demarcation of subjective and objective realms of knowledge, laid the foundations of the modern scientific age.

The mind-body problem and science itself stem from the same split in the fabric of reality. This creates fundamental problems for a science of consciousness. Science proceeds by systematic observation and experimentation. The whole point is to provide factual, public, knowledge about the world *as it is*, independent of personal feelings and opinions, stripped of subjectivity – in other words, to provide *objective knowledge*. But consciousness, in essence, is subjective and private. I can imagine your experiences, but I don't have them, and you can never have mine. Experience is a first-person business. Science operates in the third person.

'So,' I said, 'consciousness poses a forbidding challenge for science. What makes science strong as a means of understanding the outer, material world – objective, third-person observation – is precisely what makes it ineffectual when it comes to understanding the "inner world" of consciousness.'

'We can study brain states and functions,' said Number 3. 'Simply recognize that brain activity and consciousness are one and the same thing and the problem goes away.'

'Sometimes I see it that way, and sometimes I don't.'

'Because you can't make up your mind?'

'No, because there's more than one way of seeing. I agree that every conscious mental event, each and every thought and emotion, is grounded in some physical state of the brain. But there are objective, third-person descriptions of the brain and its functions; and then there are subjective, first-person experiences.'

'And never the twain shall meet? Is that what you're saying?'

'What I think I'm saying is that phenomenal consciousness – the raw *feel* of experience – is invisible to conventional scientific scrutiny and will forever remain so. It is, by definition, subjective – whereas science, by definition, adopts an objective stance. You can't be in two places at once. You either experience consciousness "from the inside" (a pang of hunger, the blueness of the sky, the chill of an autumn breeze, sunlight dappling the forest floor) or you view it "from the outside" (various configurations of neural activity and patterns of behaviour associated with different bodily states and conditions in the external environment). Science can study the neural activity, the bodily states, the environmental conditions, and the outward behaviours – including verbal behaviours that stand for different states of awareness ("That hurts"; "This tastes like chocolate"; "My heart leaps up when I behold / A rainbow in the sky . . ."), but the quality – the *feel* – of our experiences remains forever private and therefore out of bounds to scientific analysis. I can't see a way round this. Privateness is a fundamental constituent of consciousness.'

Number 2 sighed wearily. Suddenly, and uncharacteristically, I felt a surge of anger.

'And don't try to define it away!' I shouted. 'Don't tell me consciousness simply doesn't exist in the material universe – that there is just the brain and its functions – because, from where I stand, it fucking does! And, unless you're a zombie or a root vegetable, it does for you, too.'

I instantly regretted my outburst. Aware again of my nakedness, I felt ridiculous. (Never get angry with your clothes off.)

But, I thought, that's precisely the point. *From where I stand.* Only I occupy my position. Only you occupy yours.

'Might there be a convergence of the subjective and the objective if we had a detailed knowledge of our brain states, plus a more refined technical vocabulary to describe them?' It was Collicula speaking.

'No,' I said. 'Wordsworth could recast the description of his heart leaping up at the sight of a rainbow in terms of photons of refracted sunlight stimulating the cells of his retina, in turn generating specific patterns of electrochemical activity through his brain, in turn leading to stimulation of the adrenal gland, in turn causing a fluctuation in the rhythm of his heart. I am not convinced this takes us any further. It is still "his" eye, "his" brain, and "his" heart that are the focus of interest, not those of Keats or Coleridge. It is the view from where he stands. He is, essentially, irreducibly, describing a personal point of view, not a pattern of neural signals.'

'I fail to see the relevance of poetry,' said Number 2. So I quoted another poet.

'Robert Frost said that "Poetry is what is lost in translation. It is also what is lost in interpretation." Likewise, consciousness is lost in translating from first-person experience to third-person description of brain states. One can accept, as I do, that all psychological activity depends on neuronal activity, and one can chart the neural substrates of this or that psychological process, but the poetry of consciousness has been lost in the interpretation.'

'Brain activity and consciousness are one and the same thing,' said Number 3.

'Yes, there's truth in that statement.'

'Therefore, since neuroscience is best placed to describe the workings of the brain, it is clearly best placed to give an account of consciousness?'

'No. It doesn't follow.'

'I despair,' said Number 2 under her breath.

'Some people have argued that consciousness is double aspect,' I said. 'It has an objective and a subjective side. It is unique in that respect and so can't be treated in quite the same way as other natural phenomena like clouds or flowers or pebbles, which can be understood from a purely objective standpoint. There is nothing mystical about subjective reality; it is just different from the objective, science-friendly variety. It is just as real, just as material, and has nothing to do with the kind of immaterial mental stuff that Descartes believed in. Mental events are based in physical events – the two coincide perfectly. The subjective and the objective are different takes on the same underlying reality. But the subjective realm is out of bounds to science.'

'Are these your beliefs?' Number 1 asked.

Come to think of it, I really wasn't sure, and the words spilled out in the thinking: 'I'm really not sure.'

'So, what do you believe?' Collicula demanded for a second time.

I was about to repeat 'Nothing', which was, in fact, as close to the truth as anything else I might have said, but I didn't want them to think I was being perverse; and I didn't want to be there all night. So I took the easy option. I played it straight down the line.

'I am a materialist,' I said. 'I believe that the world and everything in it is made of physical stuff and, whatever the origins of
the universe, we are a natural product of its material evolution:
sentience, intellect, emotions, moral codes and all. All behaviour and experience, all knowledge and understanding of the
world and ourselves, depend upon the workings of a physical
device: the brain.'

There was a murmur of approval.

'Good. Perhaps, after all, you are not a Mysterian.'

This was a non sequitur. Mysterianism and materialism are
not mutually exclusive. But I let it pass. Perhaps I was, perhaps
I wasn't. Who cares? At any rate, the three figures seemed
happy with my pronouncement. They gathered their papers and
were gone; oddly, though, I didn't see them leave.

That could have been that; except now I found Collicula sitting astride me, her face lit by the jade waters of the aquarium
below.

'What do you really believe?' she asked.

'I meant what I said about materialism, and I meant what
I said about subjective experience being beyond the reach of
science. But, in truth, I really don't have firm beliefs on the
matter. I look at the mind-body problem one way and it seems
to evaporate. I look at it another and I'm tantalized.'

'Perhaps there's more than one problem,' she said. 'Or perhaps you are more than one person.'

I was deep inside her now and couldn't care less.

THREE

No Water, No Moon

The Ghost Tree (1)

I drive across town to the infirmary. Jake is on one of the orthopaedic wards. The beds are packed in rows along the walls. When I arrive his wife is at his bedside. She looks about seventeen, a year or two younger than Jake. There is no talk between them – a bubble of silence. I get the impression there has been no conversation for some time.

He is the image of Christ on the Cross. Matted curls of black hair drop over sunken cheeks. His forehead is bruised and scabbed where a crown of thorns might have been and a bed sheet, crumpled at his hollow midriff, serves as a loincloth. His lean, pale, upper body bears other scars of the smash: broad purple grazes and yellowing contusions. But at the bottom of the bed there is nothing. The imploding metal of the car severed one leg at the moment of collision. The other, mangled beyond redemption, was surgically removed in the hours that followed. If he is a car thief, then Jake has paid a high price for his misdemeanours. Only now, as the bandaged stump appears from under the sheet, do I notice that his right hand is also missing.

He gives a handless wave to an old man in a wheelchair.

When I pick up the medical file from the nursing station the charge nurse tells me of Jake's disturbed behaviour. Yesterday he was incontinent and smeared his faeces into his wounds. I am not looking forward to this. I feel cowardly. I want to turn and go. But when we speak he is perfectly pleasant. He seems composed, even tranquil. He puts me at my ease. The pain is tolerable now, he tells me. Yesterday it drove him mad.

'Where does it hurt?' I ask.

'Left foot,' he says.

I run through some routine questions about levels of consciousness and recall of events in the hours immediately following the accident. Jake can't remember.

'He *was* conscious,' says the child bride.

'How can you be so sure?' I ask.

She knows because Jake had activated the dial-home function on his mobile phone, perhaps adventitiously as a result of the impact or in a moment of lucidity. There was no one home when the phone rang. The answering machine took the message and stored it until she returned next morning from her night shift at the filling station. Jake was calling for her, wailing like a baby.

I do my stuff and leave. I've had enough. It is only four in the afternoon and I'm due to attend a meeting later on, but I phone in with an excuse and head for home.

It is a summer's evening, grey and overcast, perfectly still except for a tiny plane droning through low cloud, in and out of visibility. *There are people in there*, I think, but only with an effort of imagination. From this distance, who would care if it fell from the sky?

I am sitting out in the garden and I must have dozed off because the student dissertation I was reading has fallen to the ground, face down. My coffee is stone cold. Evidently, I've been asleep for some time.

I pick up the dissertation, which opens at a page containing images of a series of brain scans. I know the person whose brain this is. *MJ17* says the caption, preserving anonymity, but I know her as Maggie. She is one of my research patients. I recall the last time I saw her. It must be a couple of months ago. She greeted me, as usual, like an old friend, taking both of my hands in hers and gripping them warmly for a good minute. She took my arm as we walked down the hallway and into her living room. Then, while I'm exchanging pleasantries with her husband, Don, Maggie touches my cheek. She really has no idea who I am. Her memory is a void. This, and the lack of inhibition, is a result of the disfigurement of her brain.

There are blades of grass on the page from the freshly mown lawn and the pictures of the brain have a kind of vegetable quality. *Figure 1*, I read, *Coronal T1-weighted magnetic resonance images through the amygdaloid complex and hippocampal regions.* I am looking inside Maggie's head. She was probably humming a tune to herself as these pictures were being taken. When she is not talking she is humming or singing. Don doesn't complain.

The pictures are mostly grey. Dense material, like bone, shows up white. Darker regions signify lower density: the black butterfly of the lateral ventricles, filled with fluid rather than brain tissue; the shadowy recesses of the outer convolutions. Like a cauliflower. Large areas of the anterior temporal lobes have been eaten away by the virus. These, too, show as black.

Maggie was unlucky. The bug – herpes simplex – is very like the common cold sore virus, but it found a way into her brain. Then again, she's lucky, too. Lucky to have survived. Don thinks so. Who am I to say she isn't?

Little white arrows have been superimposed on either side of each picture to identify the regions of dark space normally occupied by the amygdala and the hippocampus: the almond and the seahorse, vital components of the machineries of memory and emotion. Their loss is what makes Maggie interesting for science.

As a clinician I have a duty to be scientifically informed and inquisitive. Someone sits before me in the clinic. They have a fault with their neural machinery and I need to appreciate its characteristics. They speak of symptoms, I listen and look for signs. I hypothesize. I test and deduce. I refer, as needs be, to the scientific literature. But I fail if, as part of this process, I do not also engage with the patient in an ordinary, human way. One has to absorb someone's personal concerns to understand their predicament. It is, after all, the *person* who is ill, not the neural machine.

This afternoon, with Jake, I had found it difficult to maintain the necessary balance between detachment and absorption. Dispassionate analysis had given way to emotional synthesis. The mutilated young man with the phantom limb, his calm civility, the devotion of his young wife, the cutting desperation of the message on the answering machine: it was too potent a mix and I was caught off-guard.

And now I seek sanctuary in the solitude of my garden and a retreat into sleep, science, and abstraction – the dissertation: the soothing icons of a bug-eaten brain. It's an effective remedy. How comforting to lose sight of Maggie and contemplate instead MJ17.

It is getting dark now. The clouds have thinned and a crescent moon is visible. At the bottom of the garden there is an apple tree. It looks tired and forlorn. This, instantly, is how I see it. It is an old tree, bearing fruit for the last time. I see not just the fading shape of the trunk, the twisting branches, the leaves darkening in the gloom and the pale, half-grown apples; I see the age of the tree and its weariness. I have in mind the sharp taste of the fruit. This is how it appears to me. And how do I know it is bearing fruit for the last time? Because I realize it is not there at all. My brain has conspired with the failing light to conjure a fleeting illusion of the tree from memories of similar grey evenings a year ago, before it was felled by a February gale. It is a ghost tree, rooted only in thought.

* * *

I am in a church. It was once a church, anyway. Now it's a university building. I'm here for a symposium and people are milling around drinking coffee before the final morning session. I keep an eye on the time because I'm presenting a paper.

The programme has reunited me with two colleagues from my postgraduate days. I haven't seen them for twenty years. We stand in a triangle. Mundane facts of biography slot together as planks in the conversational platform. We all have wives, and children, and dogs. Rick affects embarrassment. *So* bourgeois. Why haven't we had more interesting lives?

'You don't like commitment, you get married. You don't want kids, they take over your life. You get a dog, you're forever scooping shit into plastic bags.'

Steve and I concur, but we don't mean it either.

Steve has been in the United States for ten years and his voice follows mid-Atlantic contours. 'I guess the myth of romantic love is where the rot sets in,' he says, 'if you let it.' Life and relationships are more random than we think but, in the end, most of us fall into a pattern. With whom, it doesn't much matter. It's the pattern that counts. 'If you don't relinquish the myth, you're bound to be disappointed. But if you don't believe it in the first place . . .'

I climb the spiral stairway to the upper lecture theatre. The sun-filled, stained-glass window sends curves of purple, yellow, and red along the steel handrails. The hall itself is cool and dark. It fills the higher reaches of the nave. My audience trickles in. There aren't many and they scatter about the place like a congregation.

With a click of the mouse, a quotation rolls across the screen behind me: *We should take care not to make the intellect our god. It has, of course, powerful muscles, but no personality.* That was Einstein. It sets the tone of my talk, which is about how the brain generates emotions and how emotions regulate social behaviour.

There are structures for analysing the geometry of the face, and others for interpreting the meaning of expressions. These feed into systems for decoding people's intentions and dispositions, calculating their desires and beliefs. Then there are mechanisms for selecting programs of action, for shifting gear and manoeuvring the vehicle of the 'self' through the social landscape.

The amygdala is a crucial component of the social brain. It acts as a control centre linking higher cortical processes, including rational thought, to the more ancient emotional machineries lower down. In particular, it is believed to be involved in the production of fear and anger.

On the screen now is a large, moving, talking, gesturing image of Maggie (aka MJ17). She is having a good laugh with a research assistant. Two girls together. It's Maggie back in her twenties telling risqué stories about her boss. He was a *one*. Somewhere, off-camera, Don's gentle voice coaxes her back on to safer ground. He doesn't want to cause embarrassment. I've set the video in the wrong place. I intended to show Maggie and Don talking about their Spanish holiday. But at least the audience can see she is not a cabbage. She is upbeat and animated, eager for company. I'll have to tell the story myself.

They'd been out for a meal. The two boys leapt from nowhere. There was shouting and pushing. They grabbed Maggie. Don was thrown back against a wall. One hand gripped his throat, the other took his wallet. But Don is a big man. He fought back. He gave the boy a pounding. And all the time, with Don's amygdalae trilling like fire bells, jolting his body from visceral fear to thrashing, mechanical anger, pupils dilated, cardiovascular system in overdrive, blood draining from gut to straining muscles, fists like hammers – all the time Maggie smiles benignly. The fluid-filled spaces in her head where the amygdalae used to nestle are pools of tranquillity.

Back at the hotel, Don was still shaking. Maggie couldn't

fathom it. She thought the boys were just larking around.

So far, so good. No amygdala, no fear. It's a nice anecdote to colour the standard explanation of fear production: the cortex perceives, the amygdala interprets and triggers a response. But then, back home, Maggie sits watching TV. It's a soap opera. There's a spark of aggression between two female characters, nothing extreme or out of the ordinary. It gets to her, though, enough to swipe her breath, and start her heart thumping.

'No, don't,' she says. 'Please, no!'

Fear rises until the flesh of her face is pulled taut in a rictus of terror. The anecdote now becomes a window of insight into the true functions of the amygdala. At any rate, that's the way I present it. Evidently, fear can be triggered without involvement of the amygdala. Its function is to perform appraisals of danger. Maggie, minus amygdala, is oblivious to the real threat of the mugging, but shows excessive fear in response to an innocuous TV programme.

'Interesting,' someone says, 'but only anecdotal.'

I have to agree. But I'm all for anecdotes.

In his presentation Steve talks about his dog. He grants the animal a rudimentary sensory awareness, but nothing like human consciousness. His wife and kids disagree. They value emotion over intellect. They are convinced the dog has feelings – primitive and unarticulated but, at root, like ours. What perplexes Steve is that he can't help behaving as if he believes this too.

'I guess it's my social brain,' he says.

'It's a sign that you're human,' I tell him.

* * *

Maggie's story appeared in an article I wrote for a magazine. The mugging anecdote can also be found in the discussion section of a scientific paper I co-authored a few years ago, where Maggie is referred to as 'YW'. Although I think it gives an insight into the functions of the amygdala, as my critic in the audience said, the evidence is only anecdotal. But evidence like this would be hard to come by experimentally, for practical and ethical reasons. Clinical anecdotes are not only an invaluable source of inspiration for more systematic theoretical and experimental studies, they are sometimes important in their own right.

Shortly after the story appeared in the magazine I received a letter from a reader who, like Maggie, had quite recently suffered herpes simplex encephalitis. I'll call him Anthony. It was a remarkable letter. With Anthony's permission, here are some extracts:

> I continue to experience the two effects that you write about. I have both the reduced sense of personal danger, and the physical reaction to argument or conflict. On the one hand I have become a risk-taker, e.g. dangerous jay-walking (and I had a period of shoplifting), while on the other hand, I have to leave a room (escape) if anyone raises their voice, even mildly. I no longer watch TV because I cannot stand the 'tension' that stories create . . .

Anthony's combination of 'fearlessness' (or 'recklessness') on the one hand, and over-sensitivity to mild conflict and dramatic tension on the other clearly resembles some features of Maggie's behaviour. Quite likely there is a degree of overlap in their patterns of brain pathology. This would not be surprising since herpes simplex has a predilection for certain areas of the brain (the temporal and orbitofrontal regions in particular). But what distinguishes Anthony's account is how insightful and articulate it is. Maggie would have had difficulty expressing her thoughts in this way.

Then he went on to describe somewhat different symptoms of a kind I hadn't come across before.

Thankfully, some earlier symptoms that directly linked words and emotions have subsided. I used to 'feel' words. Whenever I heard or spoke a word or phrase indicating a physical state, I would automatically feel the state as well. So I know exactly what is meant by 'gut-wrenching' or 'toe-curling'. It was very disconcerting whenever people asked me whether I ever felt sad or hurt or afraid. Not only had I felt these things – who wouldn't when a virus starts invading your brain! – but I felt them equally strongly every time I was asked.

A third set of symptoms concerned Anthony's ability to communicate in face-to-face interactions with others. These symptoms are reminiscent of those reported by people with Asperger's syndrome or high-functioning autism. They are intriguing in the light of current theories about the brain disorder that may underlie autism. A number of influential scientists have implicated the amygdala.

I can no longer 'read between the lines' either, and I take people's language literally – I get little clue from their expressions. This can be hilarious, but is also very frustrating. Luckily, I am currently living with Australian friends, who value straight talking, so I have less trouble 'reading' them than I do the typical contorted English relation between words and feelings.

He elaborated on this in a subsequent communication:

Nowadays, finding it hard to distinguish levels of meaning in people's words, I am very concerned that everything be straight and true and meaningful – otherwise I do not understand. Linked to this, I will tell anybody anything – what my parents don't know about my previous sex life isn't worth knowing!

. . . The virus ate my shame.

The Ghost Tree (2)

Charles could hear the surgeons talking. One of them was angry. They were going to start cutting. There were fingers at his abdomen. Next, there would be a knife. This couldn't be. He must tell them: *Don't cut! I'm still awake! Please, not yet!*

The words formed in his brain, but their passage to the vocal apparatus was blocked. He lay motionless and mute as the blade sliced his flesh. The pain flung him from his body. Looking down on the scene from the ceiling, he saw that the angry surgeon was still complaining about something. Charles felt profound unease, not tranquillity or indifference, as some have described. How was he to get back?

The experience left him with a post-traumatic stress disorder – flashbacks, nightmares, panic attacks. Now he was seeking compensation. Intra-operative awareness is an acknowledged problem. Effective anaesthesia requires the judicious mixing and matching of drugs to patients and conditions. It is not all or nothing, like flicking a light switch. Different operative procedures demand different depths of anaesthesia, and patients vary in terms of their response. Currently, there is no wholly reliable method of detecting awareness. There are bound to be mistakes.

Perhaps one or two patients per thousand operations are able to recall events occurring during surgery. The figures are higher for obstetric and cardiac procedures, which require lighter anaesthesia. (This does not include those who may be aware at the time, but who subsequently fail to recall.) But while intra-operative awareness is a recognized complication of surgery, the out-of-body experience (OOBE) is not.

I didn't think it would help Charles's case. He'd be labelled a fantasist, which would be unfair because OOBEs, too, are relatively common – around fifteen per cent of the general popu-lation admit to having experienced one. I didn't think that Charles's soul left his body – because I don't believe in detach-able souls – but I could fully accept that he had experienced a frightening hallucination. There are many forms of intermit-tent psychosis.

I spent my first term at university lodging with a rather dour working-class family on the outskirts of Sheffield. I'll call them the Fancys, though their real name was less plausible. Mrs Fancy fed me porridge for breakfast. Sometimes I'd get back late, the worse for wear, and sometimes I didn't come back at all. I think she found me difficult. Breakfasts were bleak. We didn't have much to talk about.

Then one morning she started telling me about Aunt Judith, how she was always welcome to drop in, of course, but, dear oh dear, how she picked her times. She had turned up in the middle of the night again. Three in the morning. Third time this week.

It was tiring, especially for Mr Fancy who was on the early shift.

Aunt Judith was lonely. She would chat for an hour or so, and then she would go home. I said I hadn't been disturbed, which was true. I hadn't heard the doorbell – or perhaps she had her own key? Aunt Judith had no need of doorbell or key I was told. She had a gift. She could project her spirit. And three times that week she had projected herself through the night air to the foot of Mr and Mrs Fancy's bed. She lived in Scotland.

A day or two later Mr Fancy raised the matter again. (I wouldn't have dared.) He knew I was in on the story.

'You've heard all about our Judith, I gather. It's a bit of a nuisance,' he said, and then carried on assembling his son's train set on the flowery carpet in front of the gas fire. The four-year-old lay supine. Nothing more was said.

Just before I left the Fancys I had an unsettling experience. I woke in the early hours, aware of something glowing faintly in the corner of the room. My heart thumped an offbeat. When I turned to look, it wasn't Aunt Judith I saw but a Christmas tree. I'd got back late, let myself in, helped myself to a snack, then gone straight to bed. I hadn't noticed a tree. How could I not have noticed? I got up for a closer look. I brushed a branch and caught the scent of the pine needles.

Returning to bed I was soon asleep, but something else disturbed me. Perhaps it was voices in the street. I can't remember. But I do remember getting up to shut the window and noticing that the tree had gone. It appeared from nowhere, then, silently, it disappeared. It was there. I touched it. I could smell it.

I slept in. Winter sunshine filled the room. The Christmas tree looked splendid, red baubles and silver tinsel splintering the light. So there *was* a tree. I tried to get up, but found I was paralysed. I looked at my right arm and willed it to move. I commanded it to move. It stayed put. When I tried to sit up or roll over nothing happened. I panicked. On the inside I was a twisting fury, but the shell of my body remained motionless. I gave up the struggle, overwhelmed by an intuition that if I tried any harder I would break through the shell and float away.

I closed my eyes. The room was still a block of sunlight when I opened them again, but there was no tree.

I now recognize this as a lucid dream, an hallucinatory state in the hinterlands of slumber where the mind is alert, but the body remains bound by the paralysis of sleep – the intersection of dream life and reality. Perhaps intra-operative awareness is like this. It's happened several times since, and each time I found myself restrained by the same forceful intuition. Next time I'll grit my teeth and let go.

* * *

Not long ago I was renting a cottage on the edge of Dartmoor. It was a Sunday afternoon and I'd been working at a glass-topped table by the window. I was tired. I hadn't slept well the previous night. Now I stopped, transfixed by music.

I often work to music – usually Bach or Mozart. This was Bach; a partita for solo flute, endlessly circling and climbing, falling and rising, bright lines of sound filling the air. Great music cancels the distinction between the external world and our inner life. I was absorbed, but, also, it was me who was absorbing. I was at the centre of a machine of sound, but the machine was also within me.

I first saw the geese in reflection, through the smoked glass surface of the table, swimming in a bronze pool of sky. There were three. I looked up at the window to watch them speeding south-west under clouds that now looked unnaturally white and patches of sky unnaturally blue. I felt disembodied. It was as if I were inside the cottage, sitting at the table by the window and, at the same time, flying with the geese, high over the Devonshire woods and fields. I was dislocated and distributed, just as the geese were simultaneously in one place and another: out there, and here in the world beneath the glass-topped table.

This was not an out-of-body experience. It was not unpleasant or disturbing in any way. Subdued by fatigue, introverted by solitude, elevated by the extreme beauty of the music, my perceptions and sense of self had been momentarily reconfigured.

Our body schema is surprisingly malleable. V. S. Ramachandran and his colleagues have devised some simple exercises to illustrate this fact. I sometimes use them to enliven dull lectures. Here's an example.

First, put on a blindfold and have someone sit in front of you, facing in the same direction. Then let another person take your right hand and start tapping and stroking the nose of the person in front of you with your index finger, while at the same time using their own left hand to tap and stroke your nose. It's best if the tapping and stroking alternate in random sequences, and they must be synchronous – that is, a tap/stroke on the other person's nose must be matched by a tap/stroke on your own nose. After a while, thirty seconds or so, you may begin to feel that you are tapping your own nose at arm's length, as if, like Pinocchio's, it has grown enormous or is floating out there in front of you.

It's even possible to project sensations on to inanimate objects. Try this. You need a table and a friend. Sit with a hand under the table, hidden from view, while your friend taps/strokes the surface of the table and simultaneously taps/strokes your hidden hand. It's crucially important that you don't see what's going on under the table – that would ruin the effect, but as you watch your friend's other hand you should gradually feel the tapping and stroking sensations arising from the table itself. When it works (which isn't always), the effect is compelling. You know at a rational level that the surface of the table is beyond the boundaries of your body, but that's not the way it feels. The phenomenal experience overrides the rational analysis. The table has been temporarily incorporated into your body schema. It has become part of 'you'.

So, even on as fundamental a matter as where 'you' are in relation to your body, the conscious, reflective self is easily deceived.

∿

I liked the story of the Christmas tree, she said.

Thanks.

Why do you call it a lucid dream? Perhaps it was some other kind of vision.

'What other kind?'

Hypnagogic imagery.

Oh.

Or do I mean hypnopompic?

They're both forms of dream-like imagery at the edges of sleep, when you're dropping off or waking up. Hypnagogic is when you're dropping off. But, anyway, no, it wasn't either of those.

Don't most people fall asleep with random thoughts and pictures floating through their mind?

I suppose they do

So what's special?

Hypnagogic images are more vivid. There's more clarity and detail. They seem more autonomous as well. They have a life of their own. I used to get beautiful, weird scenes going through my sleepy head as a child – later, too, on into my teens and early twenties. It rarely happens now. It's a pity. I miss them.

What did you see?

It usually started with faces. They loomed up from nowhere. The first one always took me by surprise. They were quite ordinary, anonymous faces mostly, but sometimes they would morph into gargoyles or goblins. They seemed real. As bright as television.

Just faces?

No. Sometimes it was more elaborate – parades of little people, all bright colours like a medieval pageant. They seemed to have a life of their own. It was fascinating, and totally beyond my control. I used to watch the little people strolling

by. They always seemed to be going somewhere.
Some of them would be carrying packs or
pushing carts. I was a spectator, on the sidelines,
watching from a distance. I knew they wouldn't
bother me. I wasn't really part of it. I couldn't
enter the scene. Although sometimes it did seem
like they'd sensed my presence. One or two
would step outside the flow, come close and
look directly towards me. But their eyes were
unseeing, like I was behind a one-way screen.
I could see them, but they couldn't see me. Yet
for a moment they sensed I was there. I had no
influence over the behaviour and appearance
of these creatures, or the world they inhabited.

Where did they come from?

My brain, of course. Some hidden corner of
my mind.

Ah, but which undiscovered territory?

The fascination for me was – still is – that this
strange, nocturnal world was the product of my
brain and yet I had no conscious control over the
shape it took. I remember once looking closely at
a banner some of the little fellows were carrying.
It was beautifully embroidered, fantastic colours
– mostly reds and golds. And I thought I
couldn't possibly create something so beautiful.
I was sometimes amazed by what I saw.
It convinced me that I was just one product
of my brain's activity – a wave of conscious,
self-awareness on the surface of an ocean.

You needed convincing? I thought you were a
psychologist.

Well, obviously, I knew that a lot of mental life goes on below the level of awareness. That's orthodox cognitive psychology. And I know the Freudian stuff, and Jung. But these pictures put abstract theory in the shade. I sprang from the same source as the gargoyles, the goblins and the colourful pageants – the same brain – but I felt no connection with them.

Do you think their little lives went on when you weren't looking?

Now, that would be eerie. I'd prefer to think they didn't. I'm sure they needed an observer to bring them to life.

Perhaps we all do.

My brain conjured them up, and they required a solitary spectator – me – but once the spools were rolling I played no part. Robert Louis Stevenson had similar experiences. He put them to good use. A lot of his stories were based on dreams or hypnagogic imagery.

Which?

Jekyll and Hyde, for one.

No, which: dreams or hypnagogic imagery?

Sometimes he seems to be talking about one, and sometimes the other. From what he says about Jekyll and Hyde it was probably based on a true nightmare. But at other times he seems to be describing hypnagogic stuff. He had this technique for getting into hypnagogic states. Sometimes he would lie in bed resting his elbows on the sheets with his arms pointing upwards,

poised to drop if he nodded off. That way
he could drift into the hypnagogic world and
stay alert enough to watch the show without
dropping off completely. He talks a lot about
little people, too – the little people who run
the dream theatre.

Did they wear medieval clothes?

No, their costumes were Georgian.

And all this is different from lucid dreams?

I can't speak for others, but in my experience
lucid dreams and hypnagogic imagery are very
different. Hypnagogic images are realistic in the
way that video images are realistic. You can
observe them minutely, like when I looked close-
up at the banner. The colours were vivid. I could
see the thread. But also, like a video, you realize
you're not part of it.

And in a lucid dream you are?

For me, lucid dreams seem absolutely real.
You're right there in the thick of it, and even
though you twig at some stage and start to
appreciate that it's a dream or hallucination,
and you begin to think it through rationally –
even so, it still seems real. That Christmas tree
was there in the corner of the room as far as
I could tell. I went up close and touched it.

*Had you been overdoing the jazz cigarettes or
anything?*

No. Or anything.

Were you scared?

More perplexed than scared. Except for the feeling of paralysis at the end. That was frightening. You're stuck there powerless and you start to think anything could happen. Then there's the feeling that you might just pop out of your skin and fly out the window. I've talked to people who've had full-blown out-of-body experiences and some of them describe a whooshing, vibrating sensation just at the point of departure. I've had the same, but that's where I struggle hard to stay put. It's a long time since I had one of those dreams, but next time I really might try to let go. I doubt it, though. I'm a coward. I'm always too terrified by the thought of not getting back. Discretion is the better part of valour.

But you don't really think it's possible?

If you mean something supernatural like my soul slipping out of my skin and flying around, no, I don't think it's possible. But the thought is still terrifying. I don't believe in ghosts, but, on balance, I'd rather pitch my tent on a campsite than in a graveyard.

So you think it's possible in a different sense?

I think out-of-body experiences are real experiences, just like the phantom Christmas tree was real to me. A lot of people say they have them. But there's a natural explanation, like there is for other illusions and hallucinations.

What is it?

I don't know.

Where would you start to look for an explanation?
You could start with the physiology. There's
a pattern. It seems to happen either in states of
low arousal or very high arousal. It can happen
– probably most often does happen – just lying
in bed. But it can also happen when the person is
in mortal danger – hanging over a precipice, say.
But, for me, the first place to start looking for
explanations would be at the neuropsychological
level – analyse which brain systems might be
involved.

So, what do you think?
I think it's something to do with distortions
of body schema.

You mean body image?
Not quite. Body image is how you as a person see
yourself. It's like a mental picture you have of
your own body and it's tied up to your feelings
about it; your attitude towards it. Body schema is
more like the brain's working model of the body.

And this can go wrong?
It can go wrong in all sorts of ways. Obviously,
there's normally a tight relationship between the
body and the self. You don't get one without the
other. But in some ways the relationship is looser
than we tend to think. It's quite subtle. It isn't
that difficult to trick your brain and twist its body
schema out of shape.

*So when someone is having an out-of-body
experience the conscious, thinking part of them is
somehow dislocated from their body schema. The*

different brain systems have got temporarily decoupled.

Something like that. The body schema and the conscious self are usually in synch. But at the brain systems level they can be separated to some degree. They're dissociable.

It's plausible, I suppose. But a little prosaic, don't you think? Much more exciting to imagine disembodied sprits whizzing off to adventures on the astral plane.

Exciting, but barmy.

~

By the way, I said, who are you?

But she was already fading back into the lush darkness behind my eyelids.

The Dreams of
Robert Louis Stevenson

The past is all of one texture – whether feigned or suffered –
whether acted out in three dimensions, or only witnessed in that
small theatre of the brain which we keep brightly lighted all night
long, after the jets are down, and darkness and sleep reign
undisturbed in the remainder of the body.

Robert Louis Stevenson, 'A Chapter on Dreams'

I stood already committed to a profound duplicity of life . . .
both sides of me were in dead earnest . . .

Dr Jekyll

The Strange Case of Dr Jekyll and Mr Hyde (1886), the classic
tale of a divided self, reflects some of the moral and intellectual
preoccupations of the Victorian era: good versus evil; reason
versus passion; religion versus science; civilization versus
savagery; order versus chaos – but was also born of the double-
ness within its author, Robert Louis Stevenson.

A world traveller and adventurer, Stevenson wrote the story
in the sedate English seaside town of Bournemouth. To all

appearances he was leading the sort of life he would previously have despised: 'Respectability, dullness, and similar villas encompassed him for miles in every direction,' wrote his stepson. But the outward image of bland respectability masked the subversive machinations of his inner world; and here we have a template for *Jekyll and Hyde*.

The story is prefigured in Stevenson's earlier life and work. As a child he had been fascinated by the story of Deacon Brodie. William Brodie (accorded the title 'Deacon' as head of a guild) was a respectable Edinburgh cabinet-maker by day, but by night was the leader of a gang of thieves. He was hanged for his crimes in 1788. The story so intrigued young Louis that, at the age of fourteen, he drafted a play about Brodie. A later version, *Deacon Brodie, or the double life*, was published in 1879, and performed in Bradford three years later. Its themes were day and night, good and evil, and the duality of human personality: exposing the depravity that might lurk beneath a veneer of civility. Bolting the door and discarding his daytime garb, Brodie declares: '... by night we are our naked selves ... the day for them, the night for me.'

As a young man eager to slip the grip of Calvinistic convention in bourgeois Edinburgh, Stevenson cultivated his own, more benign, duality of character. He and his friend, Charles Baxter, 'assumed the liberating roles of Johnson and Thomson, heavy-drinking, convivial, blasphemous iconoclasts, whose sense of humour would have been a little too strong for the Stevensons' Heriot Row drawing-room'; in which guise, 'they could full-bloodedly enjoy those pleasures denied to Stevenson and Baxter, and to Dr Jekyll'. (I quote from Emma Letley's introduction to the Oxford World's Classics edition of *Dr Jekyll and Mr Hyde*.)

But there was a deeper division in Stevenson's psyche than is revealed by glimpses of his childhood obsessions, the student role-playing, and the subversive imagination. It took the form of a *dissociation*. Dissociation is a psychiatric term that refers to the splitting of mental processes from mainstream consciousness. The separated part of the mind seems to maintain a life of its own. In Stevenson's case the dissociation was evident in his dream life, and in the important part that dreams played in the creative process. He gives a vivid account of this in the essay 'A Chapter on Dreams', in which he writes about himself in the third person. In childhood he had been 'an ardent and uncomfortable dreamer. When he had a touch of fever at night, and the room swelled and shrank, and his clothes, hanging on a nail, now loomed up instant to the bigness of a church, and now drew away into a horror of infinite distance and infinite littleness, the poor soul was very well aware of what must follow ... sooner or later the night-hag would have him by the throat, and pluck him, strangling and screaming, from his sleep.'

The swelling and shrinking, looming and receding, are examples of micropsia and macropsia, pathological distortions in the perception of the size or shape of objects which come under the generic heading of 'metamorphopsias'. Micropsia refers to an illusory reduction in an object's size, macropsia to the opposite. Illusions of this sort are often reported in temporal lobe epilepsy, but may be experienced in other neurological conditions, including migraine. They can also be caused by fever, as Stevenson's account suggests, and may be quite common in childhood in the absence of illness. I certainly remember having such episodes as a young child. The description of things looming up 'to the bigness of a church' and then drawing away 'into a horror of infinite distance and infinite littleness' captures the feeling quite brilliantly.

Many of Stevenson's childhood dreams were far from fearful or disturbing. 'He would take long, uneventful journeys and see strange towns and beautiful places as he lay in bed.' The dreamer (that is, Stevenson) had 'an odd taste' for the Georgian period – consistent with his interest in Deacon Brodie – and this 'began to rule the features of his dreams . . . ' Then, as a student, he began to dream in sequence, 'and thus to lead a double life – one of the day, one of the night – one that he had every reason to believe was the true one, another that he had no means of proving false.'

One exhausting sequence of recurrent dreams was 'enough to send him, trembling for his reason, to the doors of a certain doctor'. The dream had him in a surgical theatre, 'his heart in his mouth, his teeth on edge, seeing monstrous malformations and the abhorred dexterity of surgeons'. Then he would return to his lodgings at the top of a tall building on the High Street. At least, he tried to return. Instead he found himself endlessly climbing stairs to reach the top floor, his clothes wet, all manner of people brushing past him on their way down: 'beggarly women of the street, great, weary, muddy labourers, poor scarecrows of men, pale parodies of women . . .' When, finally, he saw the light of dawn breaking through the windows he would give up the ascent, turn, and go back down to the street 'in his wet clothes, in the wet, haggard dawn, trudging to another day of monstrosities and operations'.

And then there came a turning point. He 'had long been in the custom of setting himself to sleep with tales', but, he says, these were 'irresponsible inventions'; tales told for the teller's pleasure that would not stand the scrutiny of a critical reader. They lacked all the important elements of good storytelling, such as plausible characters, a consistent structure, and a compelling plot.

His dreams, like most people's, were 'tales where a thread might be dropped, or one adventure quitted for another, on fancy's least suggestion. So that the little people who manage man's internal theatre had not as yet received a very rigorous training . . .' This is his first mention of 'the little people'.

Stevenson's dreams made wonderful raw material for his narratives and came to play an increasingly important role in his creative life. The tales began to sell and, 'Here was he, and here were the little people who did that part of his business, in quite new conditions.' A greater discipline was required. 'The stories must now be trimmed and pared and set upon all-fours, they must run from a beginning to an end and fit (after a manner) with the laws of life . . .'

Storytelling had become a business, not only for Stevenson, but also for the little people who ran the dream theatre. But, he says, they understood the change as well as he. 'When he lay down to prepare himself for sleep, he no longer sought amusement, but printable and profitable tales; and after he had dozed off in his box-seat, his little people continued their evolutions with the same mercantile designs.'

One such dream story is described at length. It is worth recounting in detail before hearing Stevenson's appraisal. He tells it 'exactly as it came to him'. The dream casts him as 'the son of a very rich and wicked man', a landowner, whom he has avoided by living abroad much of the time. On his return to England he finds that his father has taken a young wife, who is treated cruelly. For reasons not entirely clear to the dreamer, father and son agree that they should meet, but, through pride and anger, neither will condescend to visit the other; so they meet on neutral ground, 'a desolate, sandy country by the sea'. They quarrel and, 'stung by some intolerable insult', the younger man strikes the other dead.

Above suspicion for the murder, he inherits his father's estates and finds himself installed under the same roof as the widow. The two of them 'lived very much alone, as people may after a bereavement', but shared meals, spent evenings together, and gradually developed a friendship. Then the atmosphere changes. The dreamer senses that the woman harbours suspicions about his guilt. He draws back from her company 'as men draw back from a precipice suddenly discovered'. But the attraction was now so strong that 'he would drift again and again into the old intimacy, and again and again be startled back by some suggestive question or some inexplicable meaning in her eye. So they lived at cross purposes, a life full of broken dialogue, challenging glances, and suppressed passion.'

Then one day, he sees the woman slipping out of the house. He pursues her to the station and on to the train, which takes them to the seaside, where he follows her out over the sandhills, to the very site of the murder.

'There she began to grope among the bents, he watching her, flat upon his face; and presently she had something in her hand

– I cannot remember what it was, but it was deadly evidence against the dreamer – and as she held it up to look at it, perhaps from the shock of the discovery, her foot slipped, and she hung at some peril on the brink of the tall sand-wreaths. He had no thought but to spring up and rescue her; and there they stood face to face, she with that deadly matter openly in her hand – his very presence on the spot another link of proof.'

They return to the train arm in arm, journey home and settle to an ordinary evening. Conversation has been kept to the trivial. Although the woman was about to say something after her rescue, he had cut her short. Now, expecting her to denounce him at any moment, 'suspense and fear drummed in the dreamer's bosom'. But she does not denounce him. Nor does she in the days to follow. In fact, her disposition grows more kindly. In contrast, the dreamer, burdened with suspense, 'wasted away like a man with a disease'. Unable to bear it any longer he ransacks the woman's room while she is out. He discovers the damning evidence among her jewels. And, as he stands holding the object ('which was his life') in the palm of his hand, trying to fathom why she should have sought it, kept it, but never used it, the door opens and the woman enters the room.

'So, once more, they stood, eye to eye, with the evidence between them; and once more she raised to him a face brimming with some communication; and once more he shied away from speech and cut her off. But before he left the room, which he had turned upside down, he laid back his death-warrant where he had found it; and at that, her face lighted up. The next thing he heard, she was explaining to her maid, with some ingenious falsehood, the disorder of her things.'

The dream story reaches its climax the following morning at breakfast. Throughout the meal she had 'tortured him with sly allusions' but now, with the servants gone, he bursts from his reserve and confronts her. Why was she treating him so? She knew everything. Why did she not simply denounce him? Why must she torture him? He asks over and over. She, too, has sprung to her feet, pale faced.

'And when he had done, she fell upon her knees, and with outstretched hands: "Do you not understand?" she cried. "I love you!"' At this point, 'with a pang of wonder and mercantile delight, the dreamer awoke'. The story, he subsequently realized, had 'unmarketable elements', which is why he presents it to us in this brief form and didn't make more of it. But it serves to illustrate his point that the little people are 'substantive inventors and performers'.

'To the end they had kept their secret. [The dreamer] had no guess whatever at the motive of the woman – the hinge of the whole well-invented plot – until the instant of that highly dramatic declaration. It was not his tale; it was the little people's! And observe: not only was the secret kept, the story was told with really guileful craftsmanship. The conduct of both actors is (in the cant phrase) psychologically correct, and the emotion aptly graduated up to the surprising climax. I am awake now, and I know this trade; and yet I cannot better it. I am awake, and I live by this business; and yet I could not outdo – could not perhaps equal – that crafty artifice . . . by which the same situation is twice presented and the two actors twice brought face to face over the evidence, only once it is in her hand, once in his – and these in their due order, the least dramatic first. The more I think of it, the more I am moved to press upon the world my question: Who are the Little People? They are near connections of the dreamer's, beyond doubt; they . . . share plainly in his training; they have plainly learned like him to build the scheme of a considerate story and to arrange emotion in progressive order; only I think they have more talent; and one thing is beyond doubt, they can tell him a story piece by piece, like a serial, and keep him all the while in ignorance of where they aim.'

Stevenson concedes that the Little People (or 'my Brownies') do half his work for him while he sleeps. He also speculates that they might well do the rest for him as well, when he is wide awake. This is a curiously modern insight, in line with current views on the importance of unconscious processes in cognition.

'For myself – what I call I, my conscious ego, the denizen of the pineal gland unless he has changed his residence since Descartes, the man with the conscience and the variable bank-account, the man with the hat and the boots, and the privilege of voting and not carrying his candidate at the general elections – I am sometimes tempted to suppose he is no story-teller at all, but a creature as matter of fact as any cheesemonger or any cheese, and a realist bemired up to the ears in actuality; so that, by that account, the whole of my published fiction should be the single-handed product of some Brownie, some Familiar, some unseen collaborator . . .'

Stevenson wonders whether his role might best be understood as an adviser and enabler; he edits the stories; he dresses them in his finest prose; he performs the laborious task of sitting at the table and writing the words down; and he prepares and delivers the manuscript. But can he, he wonders, actually claim to be the author of the stories?

The story of Jekyll and Hyde also has its origins in a dream. Stevenson had for some time been trying to find a vehicle to explore 'that strong sense of man's double being'. As well as the play about Deacon Brodie, he had already written a story with that theme, *The Travelling Companion*, but this had been returned by an editor on the ambiguous grounds that it was a 'work of genius and indecent'. Stevenson was not happy with it either – he disagreed that it was a work of genius – and destroyed the manuscript.

Then, he says, he hit certain 'financial fluctuations', which for two days forced him to rack his brains 'for a plot of any sort' for a saleable story. And then, on the second night, he had a nightmare, screaming so loudly his wife felt she had to wake him. He was not best pleased. 'I was dreaming a fine bogey tale,' he told her. Nevertheless, he had managed to secure some key elements of the story: 'I dreamed the scene at the window, and a scene afterward split in two, in which Hyde, pursued for some crime, took the powder and underwent the change in the presence of his pursuers.' The rest of the story, he says, 'was made awake, and consciously, although I think I can trace in much of it the manner of my Brownies', adding that they 'have not a rudiment of what we call a conscience'.

Conscience makes cowards of us all.

Voodoo Child (Slight Return)

Ten minutes to go. I'm going to round off my lecture with a story. I scan the auditorium. The students are still attentive, pens and notepads at the ready. A pale girl in the front row has a small tape recorder and reaches into her bag for a replacement cassette. A pigeon settles on the sill outside one of the high windows and, watching it, I forget momentarily what I was about to say. Then it comes back to me: Robert's story.

One day, in the foothills of middle age, Robert took a long look at himself in the mirror. The reflection sent an unequivocal message. Life was running out and he was going nowhere. He was stale: bored with his job, out of love with his wife, stifled by his family, disenchanted with himself. But what struck him much harder, gripped him and shook him to the core of his being, was the thought that at the end of this dreary line of days, there was oblivion. It was time for a change.

That day on his way to work he stopped at the newsagents, as usual, to buy a newspaper. He paid for it but, on the way out, when the shopkeeper wasn't looking, Robert took a chocolate bar from a shelf and slipped it into his pocket. This little act of theft was curiously energizing. His senses felt stripped and raw and he ran back to his car in a whorl of elation. He drove faster than he should, but, instead of going to work, he travelled 320 miles from Yorkshire to Cornwall. By early evening, he found himself sitting on a beach, in the face of a warm sea breeze. Robert was profoundly happy.

The sun set, it grew dark and chilly, but he stayed there all night, conceding to sleep only as the sun rose in another part of the sky. Could he be sure it was the same sun? he wondered. He returned home late in the day with no explanation except the truth and spent another sleepless night placating his distressed wife. She demanded a more plausible version of events.

'*Robert*, what were you *thinking* of?' she said.

He said he'd been thinking about everything and had put a few things straight in his mind.

Life reverted to routine for a couple of weeks. Then, driving home from work one Friday evening, Robert switches on the car radio and hears an interview with Julian Bream, the classical guitarist. At one point the interviewer asks Bream what he thinks of 'electrically amplified guitars'. 'The electric bass is fine,' he says, but otherwise he's not impressed. What does he think of Jimi Hendrix as a player? Robert detects a note of condescension in the interviewer's voice at the mention of Hendrix, but thinks it's a good question, one he himself would have wanted to put. He waits for the reply. *Don't let me down, Julian,* he thinks. There is no let-down. 'He was brilliant,' says Bream, leaving the interviewer momentarily flummoxed. Robert gets another burst of energy like the one he had when he stole the chocolate bar.

He turns round the car, heads back into town at speed and pulls up on the pavement outside a musical instruments store. The shop is set to close in five minutes and the sales staff are cashing up. He tells them he must have a Fender Stratocaster, the guitar Hendrix played. They oblige. Robert buys an amplifier to go with it and a book containing note-by-note transcriptions of Hendrix songs. This comes to nearly a thousand pounds.

'But Robert,' says his wife when he gets home, 'you can't even play the guitar.'

He tells her he is going to learn.

But that night all elation has drained away. He lies awake until the early hours in a state of agitation, tormented by thoughts of fading into nothingness, accompanied by gut-churning feelings of the proximity of death. *Tonight, tomorrow, just around the corner. It's coming, it's coming.* He is close to panic. *It's coming, it's coming.* The next day, out of nowhere, he announces to his wife that their marriage is over and he leaves her, the house, the children, and his new guitar, never to return.

Robert goes back to Cornwall, where he finds a bar job, grows his hair, cultivates a tanned and weathered look and becomes, in effect, someone else.

Two years later, living alone in a threadbare bed-sit in the suburbs of a northern city, Robert can scarcely recollect the Cornish interlude. There are fragments, images from someone else's memory, but they don't cohere – a blue lampshade, a rainy night, the shiny, stainless-steel surfaces of a hotel kitchen, a woman (Jackie? Jenny?), a fistfight, the sea. It is hard to pull together thoughts from one minute to the next.

He feels nauseous. Something rises squirming from the pit of his stomach to his gullet. In the bathroom mirror, his reflected face seems drained of any meaning, almost the absence of a reflection. He stands staring for a while, then turns on the washbasin tap, turns it off, turns it on again, off, on, before crashing to the floor. His limbs stiffen, then jerk fiercely for several minutes, as a spreading patch of urine darkens his trouser leg. He sleeps.

This is Robert's third or fourth seizure this week. The next happens in the middle of a supermarket and, afterwards, he's taken to hospital. The doctors are concerned that, despite recovering from the fit, he has remained inert and disoriented. They investigate with head scans and find a large mass in the orbitofrontal region of the brain. It turns out to be a meningioma. This is a tumour, intrinsically benign, which has invaded the outer coverings of the brain. It has been growing for several years. By distorting the frontal lobes of Robert's brain, it was reshaping the very person he felt himself to be. They operate. Tumour excised, Robert enquires of his nurses most days: 'When are my children coming?' and 'Can I go home now?'

The lecture seems to have gone well enough. These neurogothic tales generally do. I tell them 'Robert's story' is a somewhat embellished account of a real case. I've tinkered with some of the biographical information and, of course, the patient's name was not really Robert, but the clinical details are, in essence, faithful. This man really did leave his family on an impulse following several episodes of uncharacteristically

eccentric behaviour, including acts of petty theft and sponta-
neous trips to seaside towns and other places. He really did
spend sums of money he could ill afford on luxury goods like
musical instruments (which he could not play) and expensive
clothes (which he might, or might not, wear).

He was a Jimi Hendrix fan, too. A large, iconic image of the
great man stared from his bedroom wall at the rehab unit. Hendrix, at least, remained constant in his life. Whether or not he
stood in conference with the mirror in the way I describe at the
beginning and the end, I've no idea. I threw that in. Perhaps,
somewhere, I had in mind the image of Jekyll standing before
the mirror as he watches his transformation into Hyde, and
then, at the end, perhaps it was Dracula, bereft of soul, bereft of
reflection. I don't know. It's only just occurred to me. After the
operation he really did expect to return to the bosom of his
family, unaware that they had long since moved on.

When did the slow tumour take root? How long had it been
growing and heaving its bulk into his frontal lobes, insidiously
recalibrating his personality? A meningioma like Robert's can
take years to develop, eventually becoming a stable feature of
the intracranial landscape. The brain can, up to a point, accom-
modate a slow-growing mass without betraying major clinical
signs or symptoms. It depends on the rate of growth and where
it's located. Some people grow old and die never knowing that
for half their life or more they were harbouring a benign brain
tumour. Perhaps they never know who they might have been.

I once saw a man in his seventies admitted to hospital for
investigation of a stroke. He turned out to have a tumour the
size of an orange nestling in the parietal lobe of his brain. It had

nothing to do with the stroke, had probably been there for decades and wasn't, apparently, giving him any trouble. It had become a part of him.

Perhaps Robert would have left his wife and children anyway. Perhaps he was restless and bored, or depressed. A mid-life crisis. It could be that the tumour just hastened the process or even had nothing at all to do with his impulsive decision to pack his bags and go. We can't rule this out entirely, but I think not. Impairments of social judgement, impulsive behaviour, and all the rest that emerged through Robert's personality change are a common consequence of damage to the frontal lobes.

Unlike the man with the stroke, Robert's tumour *was* causing him trouble. He developed epilepsy. But suppose he hadn't. Suppose there had been no obvious medical complications, that the tumour was just there, nudging and niggling, resetting the dials of Robert's personality. Would there have been grounds for saying that his behaviour was pathological? No. You would say it was a mid-life crisis.

Despite my undisguised haste to draw the proceedings to a close (I have a train to catch) there are several questions. Some are technical, but they are mostly about the story, *as a story*. Fair enough.

'Have you ever considered all this from a Christian perspective?' asks the pale girl at the front as, finally, I gather my notes.

'No, not really,' I say rather briskly. 'Perhaps we can discuss it next week?'

'But what happened to Robert in the end?'

'He became profoundly depressed.'

I spare her the information that after being discharged from his rehab hospital, there were two botched suicide attempts before he finally succeeded in killing himself. Third time lucky. I have this unnecessary image of Robert hanging himself with Hendrix singing 'Voodoo Child' in the background: *And if I don't meet you no more in this world / I'll meet you in the next one. / And don't be late* . . . It didn't happen that way.

My train is more than half an hour late and I kill time in a bookshop. I now regret not allowing the pale girl more time. She seemed genuinely distressed. I resolve to seek her out after the next lecture and make amends. But now I'm on the train. I have a beer in one hand and, in the other, the paperback I've just bought. It's about cosmology and I'm trying to get some imaginative purchase on the immensity of it all. It's the kind of thing I sometimes read as a way of winding down. The grandiloquent prose (*velvet mantle of the night . . . cosmic symphony of the heavens*), and the big, round numbers (*four hundred billion galaxies*) have a soothing effect.

Cosmology and neuropsychology have absurdity in common. The raw facts are strange beyond imagination.

It sets me thinking about how the physical forces that twist the galaxies and roll the train along the track connect with the social and psychological forces that animate the passengers. That recalcitrant child and his weary mother, the old couple sitting in silence, the woman opposite who catches my eye, displays a micromomentary flicker of an eyebrow and smiles as the young man with an obscene message printed on his T-shirt takes the seat beside her. Fleetingly, she and I were *complicit*. I entered her mind and she entered mine. We can plot the motions of the planets, but how do you measure the force of a glance, or the weight of a smile?

Thinking these thoughts and looking at the people around me I entertain myself by seeing them for what, at one level of description, they certainly are: complex biological machines. Physical objects. I take a little thought journey behind their eyes and all I see is darkness; then, looking to the window, against the dark, I see myself looking back at me, lost in a confusion of first and third person. The image in the window resembles a machine like the others on the train, but with an involuntary flip from third person to first, I'm back now on this side of the reflection, sitting in my own clear capsule of consciousness. I buy the illusion that other people inhabit similar capsules, but obviously they don't. And from their perspective neither do I.

I get another beer. I look again at my reflection. It chuckles. When finally I get home, I feel profoundly content, immersed in my family. Secure, immutable, invulnerable, immortal. As Robert once felt, perhaps.

* * *

The pale girl is not here today. Not in the front row, anyway. I'm early and I watch the students as they file in. The rows fill up, but she is not here. Some latecomers arrive five, ten minutes into my lecture, but she is not among them.

I pick up where we left off last week, pointing out that illness of various kinds may indirectly affect the way we see ourselves, but that neurological disease sometimes goes straight to the core and distorts the person in essence. Like parasitic wasp larvae devouring a living caterpillar from the inside, a disease can penetrate the substructures of the self – the neural systems controlling long-term memory or those that regulate emotion or the hatching of intentions or the shaping of beliefs. I remind them of Robert's slow-growing tumour and how he came to see the world in a different way. He thought differently, behaved differently, felt differently about the people around him. Was the Robert who impulsively bought expensive clothes and electric guitars, who stole chocolate bars, made impromptu trips to seaside towns, and finally walked out on his wife and children – was he the same Robert who, previously, had been so devoted to his family, worked hard to pay the bills, who would never have dreamt of stealing anything, and didn't take risks or get into fights? If not, when did Jekyll become Hyde? Was there a single incident or a single day that might be said to mark the transition? Is it possible to pin it down to a single moment? Don't we all do rash and stupid things from time to time? How many add up to a personality change?

Then the return journey. Robert's tumour was removed and he was back to something like his former self. *Something* like. He yearned for his wife and kids. He wanted them back. But in other ways he was irretrievably different, intellectually and emotionally. His mental powers were diminished. He became forgetful and couldn't concentrate. He couldn't plan things from one day to the next; his view of the future was foreshortened. His face was pressed against the wall of the present, but the past was at his shoulder. It was where he felt he belonged; in the golden valley of the time before the tumour. More than that, it was where he often believed himself to be.

This was not a wistful dwelling on the past. Sometimes he was confused to the extent that he believed nothing had changed. His wife would come to collect him. She *would*. She'd be here soon. They would pick up the children from school together. They would go home. The past was like a radio jingle; not much tone or melody, but it was in his head and would not leave him alone. Then there was the depression – and in one of these black troughs he took his own life. What relationship did post-operative Robert have to his former selves? What was his 'real' self? What was his identity?

I realize I'm waffling. Some of the students are shuffling in their seats. They have come to depend on lectures structured like self-assembly furniture manuals, with handouts and web pages full of diagrams and flow charts, bullet points and references. You give them Lego bricks of fact and opinion and you tell them precisely how they fit together. I'm thinking aloud. It disturbs them.

'Don't worry about the precise meanings of terms like *self* and *personal identity*,' I say. Ordinary language notions will do for now. Actually, I'm inclined to think that ordinary language notions are about as good as it gets when it comes to talking about 'personal identity' and the 'self', but I don't mention this. 'Think of *your* self. You know, that which you think roughly defines *you,* the conscious being sitting here in this lecture theatre; that which distinguishes you from the person sitting next to you or someone somewhere else doing something different. Or a corpse.'

A corpse? Where did *that* come from? But then an image of last night's strange dream floats before me. Matilda, one of the junior doctors, was there. We had the top half of a man's body ready for dissection. Next thing I know, the head is separated from the torso. There's Mattie, me, and some other male, I don't know who. I feel squeamish, but try not to let it show. Reluctantly, but deftly, Mattie gets started with a cranial saw.

At some level I knew it was a dream because sawing through the skull of a cadaver invariably releases the smell of burning bone – think of that acrid smell of the dentist's drill boring into your teeth. But there's no odour. No sound even. Soon the top of the skull is removed and we are looking inside at the remnants of the brain, except it looks more like a mass of melted candle wax than a brain. I can sense Mattie's disgust. *She's going to be sick,* I think. And she *is* – just a little, in the efficient, measured way that cats are sick – straight into the opened head and over the waxy brain.

This job is getting to me. Perhaps there's a part of me trying to tell me something. As if repulsed by my private thoughts (are they hovering like a polluted mist above my head?), a woman at the back of the hall stands up and makes her way to the exit.

The dream replays itself like a scene from a film. I merely observe. The macabre narrative has nothing to do with me. *I* didn't plan or construct it. It appeared fully formed in *my* dream. If someone had told me this story yesterday, as a dream vignette of their own, I would not have claimed rights of ownership. It would have seemed novel and unfamiliar. If over breakfast this morning I had been asked about my dream last night, I might well have been unable to remember. I usually can't.

The scene unfolded while the conscious, reflecting, deliber-

ating 'I' was dormant and, by the time 'I' returned to wakefulness, it had retreated into some secret compartment of my brain,
like a hermit crab folding back into its shell. Synaptically
encrypted to survive the transition between sleep and wakefulness, the virtual shell then travels with me to the university. I
bring it to the lecture. And then at some unconscious signal, or
perhaps for no reason at all, the crab emerges and the dream
story unravels in the middle of my talk. It has nothing to do
with me.

The audience settles down when I show them diagrams of
the brain and tabulate some of the clinical syndromes associated
with damage to the frontal lobes:

1. *Dysexecutive type* (dorsolateral damage); impaired judgement and difficulties with planning and problem-solving.
Lack persistence or, the opposite, persist in performing an
action well beyond the point of usefulness or appropriate-
ness ('perseveration').

2. *Disinhibited type* (orbitofrontal damage); behaviour is stim-
ulus-driven. The balance between internally generated
actions and those triggered by external objects and events is
lost. Tend to be distractible. Show impoverished social
insight.

3. *Apathetic type* (mediofrontal damage); apathy and indifference, loss of initiative, lack of spontaneity; impoverishment
of speech and thought; reduced behavioural output.

I ask them whether Robert's behaviour fits any of these schemes,
while reflecting, privately, that my teaching style today has per-

haps displayed elements of the first and second syndrome, if not the third.

In conclusion, I quickly review the main themes of the lecture and we finish ten minutes early. There are no questions. The pale girl is not here to ask whether I have considered it all from a Christian perspective. I wish she were.

This train's on time, more or less. It's seven o'clock and darker than it should be for the time of year. Suddenly I feel tired. Perhaps I'm brewing a cold. One of the students pressed a book about Buddhism into my hands as I was leaving the lecture hall. There's some stuff about suffering and death setting the co-ordinates for life. I'm not in the mood. I must have been twenty minutes on the same page.

At the station, as people are boarding the train, I watch a man and a woman on the platform. They are embracing passionately, saying their goodbyes. I'm reminded of what someone once said about partings: how the instant they're gone the person you were with seems more powerfully present than ever before. Absence is tangible. The man gets on the train, the woman remains on the platform. He looks red eyed and quite shaken. I watch the face of his girlfriend/wife/mistress as we pull away. It has a chilling composure. It is a blocking face, denying entry and exit. He won't see her again.

He sits just across the aisle from me and I feel an irrational urge to give him the book about Buddhism. I put it aside and turn to the bundle of papers I picked up at the university. I still have a pigeonhole, even though it's a year since I left. The Departmental Committee minutes are at the top of the pile but, beneath this, something catches my eye: a note about the suicide of an undergraduate. A woman, a final-year student. It is not a name I recognize. The train clatters across some points, clackety-clack, and my stomach turns.

'That girl who killed herself. What did she look like?'

I'm home. At last I've found my old address book and I'm phoning a former colleague at the university, the one who persuaded me to do the lectures. He is sluggish. It is well past midnight.

'I'm sorry,' I say. 'It's late.'

'No, it's okay.'

'What did she look like?'

'I don't know. I've no idea.'

It takes presence of mind to put an end to one's own life. Suicide may be the bitter fruit of hopelessness and despair, but it is also the end point of a decision-making process. There seems to be a 'letting go', an acceptance of the idea of death that induces clarity of thought and peace of mind. Those close to a suicide often report that the person seemed happier or more tranquil than usual in their final hours. There's something I read somewhere – I can't place it – about the causes of suicide and how they are not always obvious or predictable and how, if someone is in a particular frame of mind, it doesn't take much to tip them over – an innocent remark misinterpreted; a gesture misperceived.

I've thought about the pale girl a lot this past week, but haven't followed it up. I don't want to appear morbid or obsessive. I could have made discreet enquiries, found some pretext. It would have been a normal thing to do. But I didn't, for my own sake. I did not want to see *myself* behaving in that way, betraying signs of culpability. I'm not culpable. Yet more than once I have pictured a counter-factual world where I'm the perfect, patient teacher. 'Have you ever considered all this from a Christian perspective?' she asks, and I say: 'Tell me what you mean, exactly.' Then we have a conversation for five, ten, twenty minutes; however long it takes for me to listen to her concerns and put my own point of view gently and considerately, without crushing her. And then I would have caught my train, because it was half an hour late anyway and in my wry, atheistic way I would have construed this as a beneficent nod from the Creator, a little thanks-for-taking-the-trouble gesture.

I think about her now as I rush to my lecture. The train was late. The hall is full. They're waiting. She's waiting. There she is in the front row with her mini cassette recorder. Where have you *been*? I want to ask her. Where *were* you?

Mr Barrington's Quandary

It's Clara, my trainee, on the phone, asking me to come and see Mr Barrington. I'm forming a picture as I make my way down the corridor to the outpatient clinic. Mid-fifties, light grey suit, wet, blue eyes, sandy hair, moist handshake, the hint of a stammer. I saw him a couple of weeks ago. There before me as I enter the room is a middle-aged man, the same suit, the same eyes. But this man is completely bald. His head glistens under the strip lighting. There are tears filling his eyes and he is sweating profusely. He looks globular, dripping wet to his bones. It's a feature of his medical condition.

They had started their assessments, Clara explains, but Mr Barrington quickly became distressed and felt unable to continue. She tells me this in just those terms, as if reading from a set of notes. I make a pretence of jotting down some notes of my own, but what I have written, and am now tilting towards Clara is: *What happened to his hair?*

Mr Barrington is ahead of me. 'You're probably wondering what happened to my hair.'

Apparently it fell out at the weekend, mostly during Saturday night while he lay in bed trying to sleep. It came out in clumps as his head tossed and turned on the pillow, covering the sheets and sticking to his perspiring skin. He tried to brush it away, but the sheets were damp and he was afraid of waking his wife. Several times he went to the bathroom to dispose of the hair he had gathered, each time noticing in the mirror, without particular dismay, the virgin patches of skin advancing across his head.

'You weren't concerned?' I ask.

'No. It's the least of my worries.'

Anyway, he had lost a few clumps over the week, so it wasn't that much of a shock when the whole lot fell out. He's been under a terrible strain, he explains, and things seem to have come to a head. I note the unintended pun. Language has a life of its own. He's had these things playing on his mind, he says, this thing in particular.

'Would you like to talk about it?' I ask. 'Are you able to?'

Mr Barrington drops his face in his hands and sobs. Between bubbling sniffs and quivering exhalations he asks permission to remove his jacket. He also removes his tie. His cream shirt is marked with a bib of sweat down to the fourth button and there are large ovals of dampness under the armpits. He regains his composure and is steeling himself to say something, but is not quite ready.

'Why don't you take a break,' I say, 'get a breath of fresh air. Then, if you like, you can come back and we'll chat. We'll leave the tests for now.'

Mr Barrington just stares at the floor between us. No, he says, he must talk. It's driving him mad. But he remains hesitant. His gaze retreats to his feet. Then, looking in Clara's direction, he says if we don't mind he thinks he'd find it easier if . . .

Clara understands. 'I'll see you later,' she says, and leaves. Mr Barrington gazes out of the window across the suburbs towards the distant hills, his wet, blue eyes unblinking. He isn't admiring the view. He is adrift somewhere in a vast, inner space, the exhausted prey of a relentless emotional predator: guilt. I shake his soggy hand at the end of the session. He is very grateful. I listened. I advised. Outside it has started to rain.

Clinical supervision. While Clara fills the kettle, I think back to Mr Barrington. I see his arms swing down at his sides, his head roll back. I hear the sustained, oscillating groan like a child exhausted by a bout of crying. Then the confession: a single, weedy act of marital infidelity, a long time ago. His wife never knew. He'd almost forgotten.

'But now it's playing on your conscience?' I'd said, which was feeble in the circumstances. This was not a wasp at a picnic. It was a skewering torment.

His hair had fallen out. The storm troopers of the super-ego were doing their worst, commanding him to put the record straight with his wife. But it would break her heart, wouldn't it? What was he to do?

'Tell me what to do,' he said. 'Please.'

I wonder if we can disentangle the dilemma from the disease. The provisional diagnosis is multisystem atrophy, a degenerative condition. It carries a poor prognosis. Perhaps he's clearing the decks. But the disease is affecting his brain, so the urge to come clean, and the inability to decide what to do about it might also be understood in neurological terms.

I once made a home visit to see a head-injured patient. Someone had swung a baseball bat through the front of his skull. A
year on and he was doing as well as could be expected. He came
to greet me at the front door, but as he put his hand forward he
noticed a milk bottle on the doorstep. Before his hand connected
with mine he was bending to pick up the bottle. He had almost
reached it when he began to straighten again and turn towards
me, only to change tack and bend to the doorstep. He straightened again. He bent. He straightened. He bent. He shifted his
weight and shuffled, struggling to execute one or other of the
action plans hopelessly misfiring in the mutilated circuitry of his
frontal lobes: *motor dysexecutive syndrome*. Finally, I picked up
the bottle and gave it to him. We would have been there all day
otherwise. Perhaps Mr Barrington's quandary is a case of moral
dysexecutive syndrome.

Clara returns with mugs of tea. I feel inclined to keep Mr
Barrington's secret. He was naked enough. I won't tell her, not
yet anyway. Perhaps not at all, perhaps I'll take charge of the
case, and then she need never know.

Out of Darkness Cometh Light

Molineux. The home of Wolverhampton Wanderers. It's a long way to come for a football match. Two hundred miles and more from home. I still can't get used to the new stadium, all pale brick and mustard-coloured steel. When we arrive I feel I should be somewhere else. The Molineux of old was jagged and dark – a place of wrought iron, rough concrete, and foul smells. Now even the lavatories are spruce and well lit. A fan stands at the urinal with a pie in his free hand.

I remember the club motto – *Out of darkness cometh light* – as my sons and I ascend the steps to the Stan Cullis stand, formerly the North Bank. By the standards of Anfield's Spion Kop or the Holte End at Villa Park, the North Bank was small. From other parts of the ground it looked hunched and hooded, especially on floodlit, rainy nights. But the acoustics were demonic. The noise of the crowd was a beast. It surged up to the rafters and belted the roof like Beelzebub. At such times the North Bank was a single vocal apparatus, the crowd a steaming tongue in a black throat.

But the terraces are gone. The swirling mass of flesh is no more. Singing and chanting are more sporadic and usually fizzle out. Now we sit on plastic seats, listen to anodyne pop over the PA, watch men dressed as cartoon animals wandering the touchline, careful not to stray into opposition territory for fear of inciting the crowd. (There is hostility enough in the voices around us – this is Wolves v. West Bromwich Albion; an acrid domestic squabble.) And then Jeff Beck comes over the PA: 'Hi-Ho Silver Lining'. The crowd galvanizes. It's an anthem. I find myself singing along to the chorus: ' . . . and it's Hi-Ho WOLVERHAMPTON!' My sons look at me uncomfortably. I sing the next chorus, but with less gusto. Third time round I'm silent. I look about me and am visited by doubt. Is this Molineux? Is it me?

The other day I showed my students a video. The scene is a clinic room. A young man and an old man sit facing each other. The young man is taking a history, putting questions, carefully probing the old man's observations and recollections. The old man concentrates, giving each question careful thought, but it is clear from his responses that, despite appearances (he smiles readily, seems fully engaged and has put on a suit for the occasion), there are great voids between the sparse constellations of recollection.

He has a brain disease and can hardly carry memories from one day to the next. I made the video more than twelve years ago. The old man and my younger self are performing a familiar routine. He is dead now and it occurs to me that every molecule of my younger self has been replaced with the passage of time. In a sense, neither of those bodies has survived.

Similarly, nothing remains of the boy who stood on the terraces.

So, what survives? What makes us the same person from one year to the next, one week, one day, one minute to the next? Some philosophers have emphasized conscious recollection. Continuity of the person is down to continuity of memory. If I can reclaim the thoughts and experiences of the young clinician in the video or, further back, the boy on the terraces, then we are the same person. That's not difficult. I have clear memories of making the film and can picture the patient's wife off-camera.

My impressions of the old stadium also remain vivid. I see myself arriving, as usual, an hour before kick-off and taking my place halfway up the terraces or, when I was small, at the trench wall right behind the goal. I recall the orange gravel surrounding the pitch and the lurid green of the grass, the smells of cigarette smoke and Oxo. I have a mental image of the asymmetric outline of the stands, so clear I could draw you a picture. And, although much is a blur, I can conjure snapshots of certain games and goals. I saw these things from a particular perspective. *Mine*. I was there. It was me.

But there's a problem with this line of reasoning: amnesia. What if I *couldn't* remember these things? Would disruption of memory decouple me from the child I once was? Suppose I retained a memory link with the young clinician and that he, in turn, could recall the boy (whereas I can't). It would lead to the conclusion that the younger man and myself were the same person, that he and the boy were the same, but that the boy and I were not.

And then there's my patient. His problem was with recent memory, not remote. In all likelihood he forgot the video after a few days, but would have had no trouble reminiscing about his childhood. Was the old man fused with his child-self, but dislocated from the person he had been a week ago?

Another view is that we should abandon the idea of a persisting ego. A person is more like a club – a football club, say – existing by consensus, capable of dissolution and reconstitution. Wolverhampton Wanderers twice went into liquidation in the 1980s; the current players weren't even born when I started coming to matches; the stadium was demolished and rebuilt. Nothing tangible survives, yet here we are still – me and the Wolves.

The image of my ten-year-old self brings a churning to my chest. I feel an urge to hug my sons, but resist. They're too big and wouldn't thank me. We settle in to the match. We lose one-nil. The exit from Wolverhampton is dreary and slow, but spirits are lifting by the time we reach the motorway. We made the trip for the same fixture last season. The match video is advertised for sale on the club website. I've decided to buy it. We'll look for ourselves behind the goal, halfway up in the Stan Cullis stand. I'm going to watch me and my kids not getting any older in a universe where the score will always be Wolves 3, Albion 1. It's a restricted universe, but reliable.

To Be Two or Not to Be

He had a wild look about him. Coffee stained the front of his white lab coat, though now he was swigging water from a milk bottle.

'Call me Derek,' he said.

This was, I reckoned, the thirteenth time I'd been here. He never remembered me.

Derek was from long ago. He appeared to function now as a technician, but in the old days he was a philosopher, hauled in to advise on the metaphysics and morals of the new technology. He gave up philosophy, having solved all the problems that interested him, and now enjoyed pushing buttons for a living. As it turned out, the metaphysical implications of teleportation seemed to be no more profound than the metaphysical implications of TV. In any event, people soon got used to the idea. You stepped into the booth, you stepped out somewhere else: across the street or across the solar system.

Teleportation is speed-of-light swift. The journey to Mars, which once took several weeks by conventional spacecraft, can now be accomplished in a matter of minutes. You enter the booth and, before you know it, the door slides open and you've arrived; delivered, brisk as a blade of light, to the Martian plains. Subjectively, it's instantaneous. But, even now, many people misunderstand the basic principles. It's always been the way. How many understood the physics of TV? It's the message that matters, not the medium. So, too, with teleportation.

It does not, as some still imagine, involve breaking down the body to its constituent atoms and whizzing them off for reassembly at the destination point. What travels between the transceivers is not a stream of atoms but a stream of data. Derek pushes the green button and the scanners plot the exact co-ordinates of every atom in your body. (There are roughly ten billion billion billion of them; the devil is in the detail.) The information is encoded and transmitted from this end and received and decoded at the other, where the process of reconstruction takes place using locally available material. An atom is an atom is an atom, after all. There's nothing special about my atoms or yours. They don't carry ID labels.

One other detail: once the atomic co-ordinates have been plotted, the body is annihilated. It is instant and painless; a form of vaporization – or 'discorporation', as they call it. This happens precisely at the point of transmission. It must. The event and its timing are determined by decree of the Subcommittee on Personal Identity.

Why? Why destroy the body while the information is in transit, before the replica has been constructed? Surely, it would be better to wait those few minutes to make sure that the reassembly instructions arrive in good shape? You might think so, and the Subcommittee considered the matter carefully before arriving at a different view.

After much debate it was decided that even brief periods of 'asynchronous parallel existence' were unacceptable. In law, biological persons take precedence over their digital form. Destruction of a living body is deemed acceptable only if the digital copy represents the 'latest version'. If a person continued in biological form until his or her copy arrived on Mars (or wherever) then the replica would be 'existentially asymmetric' with the original. As the data stream traversed the inter-planetary void, the psychological life of the original would have continued to evolve. The replica would, therefore, not strictly be a replica. It would be a close match, but not exact.

To cut a long story short, it was decreed that destruction of the original in such circumstances would be tantamount to murder. The person copied must be precisely the person who arrives, down to the last atom of the last molecule of every muscle and membrane, and every last nuance of the neural nets.

My thirteenth trip. Stepping into the cubicle still gave me a tickle of excitement. I was, after all, about to be obliterated. The suspension of existence is brief. But it's real. For the duration of the transmission I would be dead, nothing and nowhere, every atom of my body returned to chaos. My heart quickened at the thought. To step into the booth was to make a leap of faith that the technology would hold good, that I would be resurrected at the other end.

Some time ago Derek placed a sign above the entrance, a fragment of an old poem: *Do not go gentle into that good night*. I noticed it as I stepped across the threshold.

Derek's smiling face appeared on a screen. 'Ready?'

I was ready. I took a deep breath. You sense a feeble, myoclonic jolt, like stumbling on the brink of sleep, and there's a momentary blackness. And that's it. Journey's end. But this time, when the door slid open, I realized I hadn't moved a millimetre.

'Problems, Derek?' I said.

He wasn't smiling now. 'Shit, shit, shit,' he mumbled.

It was some kind of malfunction. *At least I'm still here in one piece*, I thought to myself. *My atoms haven't been scattered to the ether*. Apparently, though, it was a close call (the back-up copy process had also failed), and I had to admit that I was shaken. Temporary oblivion was fine, but I wasn't prepared for the permanent option. They took me to the on-site medical facility where I strolled through a body scanner and got an unsmiling thumbs-up from the operative who then sent me on to Psychology. *Psych?* The operative shrugged: search me. *Routine*, I thought.

I read the mental hygiene posters as I sat waiting for the psychologist. She had appeared briefly to introduce herself, then left. She seemed flustered. There were raised voices somewhere. One of them was Derek's. I couldn't catch most of what they were saying, but the female voice, the psychologist's I assumed, said something about this or that being a matter for the Subcommittee. Derek said something indecipherable to which a third, masculine, voice responded: 'Out of the question!' What followed sounded like a scuffle. Next, the door burst open and there was Derek.

'Something extraordinary has happened,' he said. 'I think

you're entitled to know –' but that was all he had time to say. He was set upon by three security guards and dragged away.

I don't trust psychologists. You get the *hmm*s and the *uh-huh*s, the nods, the doggy-face expressions of concern, the unconditional positive regard, the whole professional simulacrum of empathy, and then your hour's up and they're on to the next client. It must take a certain thickness of skin or thinness of soul to do that kind of stuff day in, day out. But this psychologist was flustered, which put me at my ease immediately. She didn't know quite how to play it, so I helped her out.

'What you're telling me,' I said, 'is that I was scanned and dispatched but – obviously, since I'm here talking to you – not vaporized at the point of departure.' She nodded. 'And at the other end, meanwhile, my replica was assembled and is now fulfilling my duties on Mars?'

'That is correct.'

'Well,' I said. 'Well, fuck me. What went wrong?'

Actually, I was less taken aback than you might imagine. Human evolution has equipped the brain with an impressive range of adaptive responses for coping with all sorts of situations. It gears up the body for fight or flight in the face of physical threat, to recoil from contamination, to affiliate with its fellows, to mate and reproduce, and to come to terms with loss. But self-duplication was not a feature of Homo sapiens' environment of evolutionary adaptedness out there on the savannah. It takes a while to formulate a response.

I soon began to regard my replica as a kind of rival and wondered what it was getting up to on my behalf. In human relations similarity is often the fulcrum about which points of difference work the greatest leverage. It's what gives a personal rivalry its edge in many cases.

And what was my replica's view of events, I wanted to know. Had it expressed an opinion? At this, the psychologist's face became grave. Her mode switched from empathic therapist to purveyor-of-the-party-line. This was an exceptional event, she said. The Subcommittee on Personal Identity was meeting in emergency session at that very moment to decide the issue. What had happened was in contravention of the Proliferation of Persons Act. It was a serious matter. She was at liberty only to present me with the bare facts and regretted that she could not enter into speculation about future developments. No, I would not at this stage be permitted to contact my wife or any other members of my family, or friends, or colleagues. Or anyone. The Subcommittee was expected to deliver a statement within a day or two and until then I would be their guest. What exactly was the issue that the Subcommittee was in session to decide? She was not at liberty to say.

They took me to a small room that looked out upon a quadrangle. A solitary copper beech occupied the centre of the lawn, its purplish-brown leaves shimmering in the evening sunshine. I lay on my bunk and stared at the ceiling. It grew dark. I longed to speak to my wife and children, but communication with the outside world was forbidden. I wanted to reassure them that I was okay. I would have called them by now in ordinary circumstances. They'd be worried sick. What had they been told? And then, dropping like a forge hammer from my head to my gut, this thought: *The call has already been made.*

In time, miraculously, I slept. It was a heavy, dreamless sleep, as if I'd been drugged. Perhaps I had. But in the middle of the night I woke to find a tall figure standing at the foot of the bed. It remained motionless, its face in shadow. I would have been first to speak, had the words not lost their tired way through the somnolent circuitry of my brain.

'Listen,' said the figure. 'Listen. I think you are entitled to know.'

It was Derek. He explained, without preliminary. The Subcommittee's quandary – my mortal problem – was this: given the unfortunate turn of events, they were now debating whether to allow me and my replica to continue to exist in parallel, and thereby contravene the Proliferation of Persons Act, or to have one of us, even at this late stage, vaporized. To be two, or not to be. It was a hard one to call, he said. In law, the creation of surplus individuals was a serious crime; the mirror image of murder. To Derek's knowledge, my replica had not immediately been informed of the circumstances, so had carried on as if nothing untoward had happened. He could not say for sure that it had been informed even now. This could be an influential factor if the Subcommittee decided that one of us had to go.

If the replica's relationship with my wife had evolved even to the merest extent of a brief televisual communication then that could weight the decision in favour of allowing the replica to survive rather than me. Discorporation was not my preferred option. But surely, I thought, the proposal would not be carried anyway. They *couldn't*. It was preposterous. Having presented itself, the dread proposition had to crank through the archaic, clockwork logic of the committee process, but then, *surely*, it would be thrown out.

'I don't want to die,' I told Derek.

'It wouldn't be the end of the world,' he replied, which struck me as an odd thing to say.

Derek sat in the easy chair at the far end of the room, still in shadow. 'What's the difference?' he said. 'Suppose it had gone according to plan. You would have stepped into the booth, the scanners would have done their stuff, your body would have been zapped to zero, and your replica would have appeared on Mars, walking your walk and talking your talk. And that is what has happened – that is what always happens – except this time the zapping may have been a little delayed.'

The difference, Derek, I might have said, is that I'm still here, *now*, and, having had time to reflect, I don't think I want to be zapped to zero, even though, twelve times before, this is precisely what has happened.

Nevertheless, Derek had a point. Each of the previous times I'd been teleported to Mars the experience was the same. I walked out of the booth with perfect recollection of the day's events up to the point of standing in the transceiver on Earth and experiencing that familiar little jolt and the brief blackness, and then there I was taking in the Martian landscape. There was perfect continuity. The twelfth time I remembered the eleventh, the eleventh time the tenth, and so on. And each time I could reflect back not just on that day's events, but on events of the previous day, too, the previous weeks and months and all the years of my sentient, self-conscious life.

On arrival, I always called my wife, told her I was okay, that I missed her already, and checked on the kids. Then I went about my business. And when I slept I knew my dreams had made their digital way across the void with the rest of me. They had their familiar fabric, the usual blend of the mundane and the mysterious. I dreamt of home, of work, of ordinary things. And then it was no surprise to meet the lost and the dead and run unfettered by logic and time through the streets of my childhood or take wing over oceans and strange cities; I dreamt secret dreams. This is what had always impressed me most about teleportation. Not only was the body reconstructed in perfect replica, and the conscious mind, but the unconscious mind, too: those things hidden from the observing 'I'.

Now my replica was dreaming those dreams, and before it slept it had called my wife, told her it missed her already, checked on the kids, and gone about its business. *It* had done those things, not I. We were not the same. I was flowing in a different stream of consciousness. But if the replica wasn't me now, how could it have been me on the previous twelve occasions? What did the experience of perfect continuity amount to? Was it no more than the illusion of life disguising a dozen deaths?

It was just three weeks since I'd last made the trip. Did that mean that, as a sentient, self-conscious being, I was less than a month old, exquisitely configured from the chaos of a billion billion billion atoms and artificially equipped with the memory banks and dispositions of a middle-aged man? If so, my identity was a fiction.

'The problem,' said Derek, 'is that most of us have false beliefs about our own nature. People expect determinate answers to questions about personal identity: "Yes, it is the same person" or "No, it isn't." That's one great misconception. The other is that personal identity matters in the first place.'

I experienced a slow infusion of anger, rising in my chest, diffusing to my face and fists.

'It's fine for you, Derek,' I said, 'to deny the importance of personal identity and pontificate on the conceptual confusion of anyone else on the planet who happens to believe otherwise. But put yourself in my position. There's a distinct possibility that I'm about to be snuffed out. Right now my concerns about whether I shall still be here by the weekend – or "zapped to zero" as you indelicately put it – seem real enough. And if I am to be vaporized, I'm sure you'll have no difficulty giving a determinate answer to the question of whether I exist or not.'

'Well,' he said, 'that's not quite the point.'

I'd been sitting on the edge of the bunk, but stood now and moved towards him. Both of my fists, I noticed, were tightly clenched.

'Derek,' I said, 'you'd better go,' at which he raised a placatory palm, acknowledged my distress, and said he was here to help. In fact, he'd been through something similar in the early days, since when he had achieved a kind of insight. His travels in philosophy and daily exposure to the plain facts of teleportation had brought him to a vision of the self, which, once absorbed, began at once to draw the sting of death. This most natural of fears was revealed as synthetic. It could be dismantled. Intellectually.

He had once watched himself die, he told me. It was one of the first interplanetary teleportations. His first, and only, visit to Mars. He entered the booth and followed the usual procedures and, sure enough, stepped out into the reception zone at the Martian base as if he were stepping out of his front door. It was a big event in those days. They were ready with champagne and smoked salmon to celebrate. At first no one in the reception party was aware of the malfunction back on Earth. But then the message came through. Scanning and transmission had worked a treat – of course they had, there he was, soaking his reconstituted flesh in champagne – but the vaporization phase had failed to kick in.

Derek had arrived but, at the same time, he hadn't left. And was it for better or worse that the Earthbound version of Derek had suffered fatal injuries in the process? The discorporation mechanism had stuttered and stopped. His whole body blinked on the brink of extinction, fading then regaining its shape, but only at the cost of significant damage to the cardiovascular system. He would be dead within a week.

Derek 2 had returned at once, not knowing what to expect.

'I tried to console him,' he said. 'I told him I loved his wife; I would care for his children; I would finish the book he was writing. And, of course, from my perspective, nothing had changed: they were my wife and children, it was my book and it was my intention that I should finish it. So, I told him: "Don't despair; nothing will really change." But he wept. He said that no doubt I would do all those things as well as he could, and it was some consolation that his family would not suffer the pain of bereavement, but the fact remained that within a few days he would lose consciousness for ever. This would be a terrible loss.

'He had been thinking of a home movie, the one where his smiling daughter – my smiling daughter – is standing in the kitchen with a basket of strawberries. She's about three years old. There is sunlight streaming through the window on to her face. And she takes the fattest strawberry in the basket and crams it into her mouth; it's so big she has to push it in with the palm of her hand. Her cheeks are bulging as she struggles to chew. Her eyes are closed. She is utterly absorbed, over-whelmed, by the experience of the fruit. She even sways a little from side to side, as virtuoso violinists do. "That's what it's all about," he said. "Conscious experience. And that is what I shall lose; that beautiful smile, the taste of strawberries, fond memories." And to that extent, he was right,' said Derek. 'His consciousness would fade, beyond darkness and silence to oblivion. He would become nothing.

'I was with him when he died. There was no one else around. We had agreed that it would be in the best interests of the family that they should never know of our duplication. Why should they be troubled? There was no need. Let life go on as normal. I admired his resolve at the end. It made me feel proud. He so wanted to see his loved ones for one last time, but understood the distress and confusion it would cause. So it was just me and him. I held his hand. And then life did go on as normal. I went home and hugged my wife and children, and eventually I finished my book.'

'Then it can't be denied that personal identity really does matter,' I said. 'Your former self died a lonely death. His con-sciousness switched off like a light. He lost everything: beautiful memories, the love of his family, hopes and plans for a future that he would never reach; life itself. Those things made up his identity. Nothing more mattered, and nothing mattered more.'

Derek leaned forward, elbows on knees, thumbs to cheek-bones, fingers to forehead. 'Yes and no,' he said.

We talked for hours, until the light of a grey dawn conjured shapes in the courtyard as if from imagination. The copper beech seemed reluctant to appear. Was this to be my last day?

Derek did most of the talking. He must have thought these things through a thousand times before, but still there was a note of urgency in his voice, as if he were on the brink of a revelation. I am not a philosopher, and at times I found him hard to follow, but I got the gist.

He explained that there were two ways of looking at a person or, rather, two theories about what persons are, and what is involved in their continued existence over time. The first theory he called Ego Theory. This is the intuitive, common-sense view, but one that was held, also, by some of the greatest philosophers, most famously René Descartes. It made sense to me, too: I wake up in the morning; I go to work; I feel happy when things go well and I feel frustrated when they don't; I hold certain beliefs about the world and express various opinions and prefer-ences: I used to like Beethoven, but now I prefer Mozart; I like chocolate better than cheesecake; I enjoy walks in the country-side; I take the view that people should be kind to one another, and I feel bad if I do the wrong thing. I act, I feel, I think, I believe, I grow older, and I change in other ways. But 'I' am always there at the centre of things as time goes by.

What is this 'I'? We ordinarily claim ownership of our actions and thoughts and experiences: *I* did it; that's *my* idea; *I* feel hungry; *I* intend to buy a birthday present for *my* daughter

. . . So the 'I' is the experiencer of experiences, the thinker of thoughts, and the doer of deeds. Each day is a blizzard of sensations and thought patterns, but *I* give them coherence and link them to *my* memories and *my* plans for the future.

It's natural to think in this way. We are the progenitors of thoughts and actions and they are *ours* in the thinking and doing. According to Ego Theory, it is this 'I' that constitutes the essence of the person and which persists over time. But, again, what is it? Descartes believed that the ego was a purely mental thing, a soul or spiritual substance, but you don't have to go along with that to accept the idea of the self, the ego, as a kind of hub about which the wheel of experience revolves.

In this non-spiritual sense the ego is merely the *subject of experience*, it is that which unifies someone's consciousness at any given moment. I watch the sky lighten, I see the leaves of the copper beech gain colour, and I hear birds singing. What gives this scene its unity? What pulls these disparate threads of experience together? I do. They are experiences had by me, *this* person, at *this* time. And the wheel rolls on through the years, accounting for the unity of my life.

These thoughts were running through my mind as Derek spoke. *I* was thinking them. 'I can seeing nothing to disagree with there,' I told him.

Derek stopped talking and, for a moment, there was only birdsong; then he turned to Bundle Theory. He said that like many styles in art – such as Gothic, baroque, and rococo – Bundle Theory owed its name to its critics. But the name was good enough. This theory rejects the idea that actions and experiences are owned by some inner essence, ego or 'I'. There are just sequences of actions and experiences. Nothing more. Each life is a long series – or bundle – of mental states and events, bound together by various kinds of causal relation, such as those linking the perception of a fierce-looking dog with the emotion of fear and the disposition to run away, or the different causal relations that hold between episodes of experience and episodes of memory. And that's all. The idea of a central ego, or person, contributes nothing to our understanding of the unity of consciousness at any given time, nor does it pull the golden thread of experience through a lifetime.

'So,' said Derek, 'from this perspective the ego is a hollow fabrication, and you could even say that Bundle Theorists deny the existence of people.'

'But that's absurd.'

'Yes,' he said, 'and you are not the first to say so. The eighteenth-century philosopher, Thomas Reid [he came to be known as 'the common-sense philosopher'] made a similar objection. "I am not thought," said Reid, "I am not action, I am not feeling; I am something which thinks and acts and feels." Yes, of course, that's what it seems like for all of us, and it's certainly the way we are used to talking about ourselves and others – as if there really were some central nucleus of a self, a ghostly pilot setting the course and handling the controls. "Don't call me a sequence of events," you say, "I am a person, a *person*, a PERSON!"' Derek beat his fists on the arms of the chair for emphasis.

'Fine,' he said. 'Bundle Theorists accept this as a fact. But they accept it only as a fact of grammar. People and subjects-of-experience exist as a feature of the way we use our language, but in no other way. If you say there is more to it than this, if you say there is something behind the chains of interacting mental events and brain functions, something above and beyond, observing and controlling, bundling it all together, holding its shape from one day to the next, then the Bundle Theorist would say that you were profoundly mistaken.'

'I think you're beginning to lose me now,' I admitted. 'I can accept that there are many processes going on in my brain of which I am unaware, all sorts of hidden machineries producing thoughts and perceptions, shaping speech-patterns, influencing decisions and actions in ways too rapid or subtle to be picked out by the spotlight of consciousness. But, once such things are in the spotlight, who or what is having the experiences, if not I?'

The sense that I was author of my own thoughts and actions felt like more than a 'fact of grammar' to me. Derek merely replied that yes, it was indeed difficult to accept the truth of the matter. He said there was a conflict between scientific understanding and people's ordinary intuitions about what they believe themselves to be, because there is nothing in the brain sciences to support Ego Theory.

Few, if any, neuroscientists believe that there is anything corresponding to a self or ego distinct from a multiplicity of mental states and their associated patterns of brain activity. From the perspective of neuroscience, Bundle Theory is obviously true. But Ego Theory won't go away. We can't shake it off. The beliefs that most of us hold about our continued existence over time are built upon assumptions that Ego Theory, or something very like it, is true.

'That's what I meant,' said Derek, 'when I said that most people hold false beliefs about themselves.'

Bundle Theory was not a new idea, he explained, just a difficult one to come to terms with. Its roots reached down to the sixth century BC and the teachings of Siddhartha Gotama, the Buddha, 'the enlightened one'. *Anattavada*, the Buddhist doctrine of 'no soul' or 'no self', holds that people and selves have only nominal existence (as opposed to actual existence), meaning they are just combinations of other elements. The self is no more than a bundle of fleeting impressions.

Derek quoted from memory a segment of some Buddhist text: '*A sentient being does exist, you think, O Mara? You are misled by a false conception. This bundle of elements is void of Self. In it there is no sentient being. Just as a set of wooden parts receives the name of carriage, so do we give to elements the name of fancied being.*'

'Now,' he continued, 'when teleportation came along many people had grave misgivings. They saw it not as the fastest means of transport, but as a sure means of dying. True, if you submit to the process, your replica turns out perfect in every way, with an identical body and brain and identical patterns of mental activity, including memory systems replete to the last atom and iota of information. "But," they said, "*don't be fooled.* Though it might resemble you in every way, the replica will not in fact *be* you. It will be someone else. It can't possibly be you because your body and brain have been destroyed."'

'And they were right,' I said. 'My present predicament proves it.'

'Perhaps,' said Derek. 'In a way. But not in any way that

really matters in ordinary life. Not if Bundle Theory is true, as I believe it to be. The fact is that teleportation became commonplace. It became a tried and trusted mode of transport and no one had any complaints. People went into the booth and they came out at the other end, intact in body and mind. Life went on as usual. You've done it numerous times yourself and it's never been a problem, at least not until now. And I want to persuade you that, even now, even if the Subcommittee comes to the conclusion that you are to be vaporized, it isn't really as much of a problem as you fear.

'Let me put it this way. Even though it involves destruction of the body and reconstruction using entirely new materials, we should think of travelling by teleportation as no more threatening or problematic than travelling on life's journey from one day to the next. What matters in both cases, in terms of what is preserved, is precisely the same: psychological continuity. We are the same from one day to the next only in so far as the bundle of mental states, actual and potential, that our brain takes with it to sleep at night resembles the bundle that it wakes up with in the morning. You survive from one day to the next because the psychological links have been maintained.

'On Tuesday you have a certain set of memories and plans, aptitudes and dispositions. These flow from the ones you had on Monday and are, in turn, causally linked to the ones you will have on Wednesday, Thursday, and Friday. And if, on Saturday, you are teleported to Mars, your replica emerges with the very same pattern of mental states, which will be carried forward to the next day and the next and the next through the usual causal links. There is no break in the continuity of mental life. You and your replica are psychologically continuous at the deepest level. And you must realize that the mechanisms of mental survival over time in ordinary life really are no different, and that there is no other kind of continuity that really matters. There is no point peering into the bundle, hoping to catch a glimpse of some elusive, observing ego. There isn't one. The bundle is all.'

I was beginning to understand, but it didn't help my case. I might still have to face the prospect of an untimely death. I said that Bundle Theory might very well be true, as Derek believed, and my mind would live on in replicated form – and, yes, there was some consolation in that. The replica moves forward in time with my stock of memories and dispositions. It can go on to fulfil my plans and obligations. Perhaps it really was the case that, by any meaningful analysis of the nature of mental life, I stood in relation to my replica as I stood in relation to the person I was yesterday and the person I might be tomorrow. But, at the same time, it was also clear that a branching had taken place.

While the replica's mind had rolled out with perfect continuity from the mind I embodied at the point when Derek pushed the green button to initiate the scanning process, our minds had, since then, begun to diverge as, minute by minute, we absorbed different experiences. We did not know whether the replica had even been informed of the teleportation malfunction. If not, then it would be carrying on as normal – as me – happily unaware that a version of its former self was languishing miserably on a truncated branch line.

Would it care? I wondered. I liked to think that I (and therefore it) would feel compassion. But really who knows how one might react if some bizarre alternative version of one's self showed up like the spectre at the feast, threatening to warp the status quo? It could prove a dreadful encumbrance.

I put it to Derek that my replica and I may once have been identical, but we were growing apart. We were developing alternative points of view. We had become different people if only in the restricted, language-dependent sense that Bundle Theorists allowed talk of people. And I, from my perspective, did not relish the prospect of a premature death, notwithstanding the considerable difficulties that future life with a duplicate would inevitably entail. (As the night wore on I had been thinking more and more about my wife and what her strategy might be for coping with a duplicated husband.)

'I can understand,' Derek said, 'that you don't want to die. Even Bundle Theorists don't want to die. For most people the truth of Bundle Theory does not dispel the illusion of the ego. They cling to their false beliefs.' But he also said that acceptance of that truth could have a liberating effect. Among other things, it opened the possibility of looking at death in a different way. He personally had found this to be the case. It was liberating and consoling. Before he had fully absorbed the truth of Bundle Theory he said that he had felt imprisoned in himself. His life seemed like a glass tunnel through which he was moving faster and faster every year and, at the end of which, there was nothing but darkness.

'Now my view has changed. The walls of the glass tunnel have dissolved. I live in the open air.' It had brought him closer to people. He was less concerned about his own life and more concerned with the lives of others. He cared less about his own inevitable death. Mental events and experiences would continue to be a feature of the world after his death, he said, and it was true that none of these would be linked to his present mental life by the sort of direct experience-memory or intention-action connections that shaped the existing bundle of experiences. There would be some indirect connections, however – memories of him, thoughts influenced by his thinking, advice followed.

'We all make a mark,' he said. 'The simple facts of death are this: that it will break the more direct relations between my present mental events and mental events arising in the future, and that certain other relations will not be broken. Once the ego is removed from the scene, that is all there is to it.'

I found this more depressing than uplifting. I told him that his redescription of death seemed to me to reflect an impoverished view of life. Perhaps it was better to anticipate losing the self in death than to deny it in life, a denial that, surely, amounted to a form of nihilism. And, anyway, even if Bundle Theory was true I could not believe it. Intellectually, I could follow his arguments and accept the facts of neuroscience but, psychologically, it was impossible to identify with his theory. It ran counter to one's experience of the world.

Derek's response, again, was that it was difficult to grasp the truth, and if he was being charged with nihilism then he would accept that in so far as it applied to his view of the self. The term, he reminded me, is from the Latin *nihil*, meaning 'nothing'. 'It is perfectly true to say that the self is No Thing.' Otherwise, he rejected the charge. He said that in accepting the truth of Bundle Theory his appreciation of the value of life had only been enriched.

Apparently, in finding Bundle Theory depressing and unpalatable I was in good company. Derek said that the philosopher David Hume, who had formulated an influential version of the theory in the eighteenth century, reflected on his own arguments and was pitched into a deep depression, the cure for which was to get out more – dining and playing backgammon with friends. And in the twentieth century, Thomas Nagel also came to the conclusion that, whatever the truth of the theory, it was impossible for the human psyche to digest.

'Derek,' I said, 'perhaps you should get out more, too. Rediscover your self.' He rose from his seat, grinning broadly, and came across and slapped me on the back.

'Maybe,' he said. 'Maybe not.' Then he wished me luck and was gone.

I dozed off and dreamt I was in the beautiful city of Venice, in the Piazza San Marco. Bells were ringing and pigeons circled and swooped. There were no people. Then, across the square from the direction of the Basilica, there appeared a young couple. They walked towards me. The woman I recognized as my wife, a younger version, perhaps as she would have been when we first visited the city. The man had no face, just a smooth plane of skin where the usual contours and orifices should have been. There was no greeting.

I woke to find that a white envelope had been pushed under the door. It contained an invitation to appear before the Subcommittee on Personal Identity at eleven o'clock that morning. It was now ten. I showered, breakfasted on bread, cheese, and coffee, and prepared to meet my fate.

Gulls

'I've got a lump,' my wife says. 'Feel.'

I'm watching the Sunday afternoon football and still have an eye on the game as she directs my hand to the outer curve of her left breast. It is the sixty-seventh minute. There is a lump.

'What does it feel like?' We are in the bedroom now. She has stripped to the waist. I palpate the breast as if I'm an expert. 'What do you think?' she says.

'There's a bump. More of a bump than a lump.'

The GP is reassuring at first, but develops doubts, and organizes a referral to the breast clinic. Kate returns tearful. We both need cheering up, so we head for the coast, half-an-hour's drive, stopping to buy sandwiches on the way. It's a fine day, as bright as the calling gulls. The tide is low and the sea is a hard blue. It feels good to be alive. But later, in the pub, it's my turn to be tearful. The second beer helps.

Thursday week, the day of the appointment. We sit in the grey-blue waiting room at the breast clinic. There is a TV in the corner with the sound muted. Some TV chef is baking a birth-day cake. No one is interested.

Our turn. Through to the examination room, my semi-naked wife looking fragile as the surgeon prods and palpates and gets to work with his blue marker pen. He doesn't say much. The nurse does most of the talking. She's lovely.

Our turn again. Yes, there is something suspicious on the X-ray, the surgeon says. He's going to do a core biopsy. Seven slamming shots of the silver gun, and each time he holds up the phial to inspect the maggoty plugs of flesh. He's not quite get-ting what he wants. We'll have the results in a few days.

'But you don't like the look of it?' I ask.

'No.'

'You think it's cancer.'

'Yes.'

The nurse specialist has joined us: bad news personified. The surgeon leaves and it's the three of us in the examination room, Kate's tears hot on my shoulder. The nurse sits quietly. I have my back to her, which feels like a discourtesy.

It's a nasty, sticky word, 'mastectomy'. I don't like the sound of it coming from the surgeon's mouth. We are back in the consulting room. He's plotting the likely course: surgery, chemotherapy, and radiotherapy, not necessarily in that order. We don't go straight home. We stop by the riverside and walk in the woods. I can't remember the last time I wept.

A week on. They're setting it up as a Bad News Consultation. They have grave-jolly faces – professional wistfulness. But we know already. There isn't much by way of preliminary. The surgeon squints over his spectacles.

'It's a malignant, invasive, ductal carcinoma,' he says, with a trace of apology. Is there anything else we want to know? He's not going to tell us unless we ask.

'What about the histology? Are the cells well, or poorly, differentiated?'

'Poorly,' he says, 'Grade 3.'

It's a bad one. He replays the likely treatment scenario: four cycles of neoadjuvant chemotherapy over three months; mastectomy, followed by radiotherapy. The plan is provisional, because if the chemo fails to shrink the tumour they'll bring the surgery forward. Arrangements will be made for bone and liver scans. We can do bloods and a chest X-ray straight away. I keep saying 'we'.

How odd this is. The worst news, but a sense of relief. I have already pictured the surgeon slicing off my wife's breast. I have imagined it being thrown to waste. I have seen it rising in smoke through the incinerator chimney. Yet there is comfort in the thought of getting on with treatment. Whatever it takes. It almost feels relaxing to walk out across the main concourse of the hospital – like a departure lounge with its café and shops – this place I know in a parallel professional life, out into the sunlight.

This time we don't weep by the riverside or walk in the woods. We head for the supermarket. A familiar-looking man in shorts and T-shirt is loading his shopping into the back of a Volvo. It's a famous TV newsreader. I want to tell him our news. Back home we drink beer and eat curry and watch football.

We are handed over to another surgeon, a specialist in breast reconstruction. Kate sits on the edge of the bed. The surgeon stands stroking his chin, observing her bust with the eye of a sculptor. He stoops, presses and probes the diseased breast, then stands back for a fresh view. He takes out a little ruler and starts measuring. He's weighing things up; thinking on his feet. Yes, he says, we could go for a wide local excision instead of mastectomy. And now, despite myself, I find I'm playing Devil's Advocate. I've read the latest *New England Journal of Medicine* and understand that, all things being equal, less radical interventions are just as effective. But I need to hear it from the man in the dark suit.

Five months on, post-chemo, Kate lies on a hospital bed, draped in drips and drains, recovering from her second operation in a month. The first was to remove the lump. This one has restored the breast to its original shape, though we have yet to see the sculptor's handiwork. She is swaddled in a 'bear hugger' blanket, filled with warm air to aid the perfusion of blood. With her hair just starting to grow back she looks like the Dalai Lama, but much prettier.

* * *

The year before last. Evening on the terrace of a French seaside hotel, late summer, the sea as smooth as mercury, the sky not yet drained of its blue. There would be stars, but time had slowed. Even the gulls seemed suspended on the cooling air, and made hardly a sound. They are more soft-spoken here. Kate and I were drinking cold beer, recovering from the heat of the day, our skin feeling full of the sun, our limbs aching from a long swim. We said little, but sat content watching the darkness gather. The candle on the table remained unlit.

I became aware of the man and woman two tables along. She had said something inaudible, to which he had replied '*Non, merci*,' but nothing else was said.

The man, about forty, sat hunched with arms folded, as if constrained by a straitjacket. His face had a drawn, intent look. He could have been concentrating hard. From time to time, his lips pursed and his right shoulder seemed to jerk forward a little. I watched, discreetly. I noticed, too, the squirming movement of his right hand. Pressed between forearm and biceps of the left limb, it was trying to escape.

A waiter appeared from nowhere offering something, but the woman waved him away. Kate had her back to all this and couldn't see what was going on.

Somehow, the man and woman on the terrace brought to mind a scene from a novel I had been reading: Kundera's *Immortality*. One of the characters, Agnes, is lying in bed next to her husband, Paul. Both have difficulty sleeping and Agnes drifts into a familiar fantasy about a kindly visitor from another planet. The stranger tells her that in the next life she will not be returning to Earth.

'And Paul?' she enquires. No, she is told, Paul won't be staying either. What the stranger needs to know is, do they want to stay together in the life to come or never meet again? In Paul's presence, Agnes always knew she would be incapable of saying that she no longer wanted to be with him. How could she? Wouldn't that amount to saying that there had never been any love between them, that their life had been based on the illusion of love? And for this reason she would always capitulate. Against her wishes she would tell the stranger that of course they wanted to remain together in the next life.

I put the same question to my wife. 'Imagine,' I said, 'a visitor from outer space joins us now at this table. He makes his offer of another life beyond death, and gives you a choice. You can make arrangements for me to join you or you can decide that, at the end of this life, we should part company never to meet again.'

The hand escapes. It writhes from under the left forearm and pushes forward, palm facing outward to the sea. The man's expression does not change, he looks straight ahead, but I see that his knees are now pressed tight together and are edging to the left as his upper body twists to the right. The woman takes the errant hand and puts it between her palms. She guides it back towards the man's chest. She holds it fast with her left hand and, at the same time, reaches for his left with her right. Still he looks straight ahead, does not speak. I watch as she refolds the arms, pulling them tight like a knot. The man gives no acknowledgement. She returns to her seat.

Kate was still thinking about the question. Too long, I thought. I wouldn't believe her now. Then, straight out, she said: 'I'd go it alone. Wouldn't you?' She said that one lifetime was enough, however much you loved someone.

Two more couples came to sit at the table between us and the French couple. I'd seen them from a distance at the beach that afternoon. At first I'd assumed they were French, but there was a self-deprecating jokiness about one of the women as she struggled, inelegantly, to get into a wetsuit. I thought it betrayed her as English. I was right. They ordered a round of drinks, then another, and another. Their candle was lit, and it lit the reddening faces around the table and guttered in the glass of the accumulating bottles.

The French couple were away in the shadows. You could hardly see them now. But a second and third time I noticed the man's hand escape and, each time, saw the woman retrieve it. He jerked and writhed as she tied the arms together, before settling back into a bunched repose, set to unravel again at any moment.

It's Huntington's disease, I thought. *St Vitus's dance. Poor man. Poor woman.*

The involuntary, choreiform movements are only the half of it. There's the dementia, too, and perhaps psychosis. The disease is relentless and he will dance like a puppet to his death. His fate was fixed at conception. A rogue gene, dormant for decades, had struck, and his brain was crumbling at its core, deep beneath the wrinkled mantle of the cortex, down in the dark interiors of the basal ganglia, where actions are deciphered from the codes of intention. And now all is confusion. There are actions and intentions, but they don't necessarily coincide. Meanwhile, they will do their best, this couple, to deny the terror and defy the puppeteer. They will enjoy a summer's evening together out on the terrace.

I absorbed my wife's answer – *I'd go it alone. Wouldn't you?* – and took another sip of beer.

'Would you?' I asked. Someone in the English group knocked an empty bottle off the table. It shattered on the ground, bursting shards in every direction. I noticed that the French couple were gone.

'So,' I said, 'it's just the one lifetime?'

'Afraid so,' she said. 'Better make the most of it.'

Remember me when I am gone away,
  Gone far away into the silent land;
  When you can no more hold me by the hand,
Nor I half turn to go yet turning stay.

Christina Rossetti, 'Remember'

FURTHER READING

Some of the following readings relate to particular chapters in this book, others will illuminate its general themes. Still others are recommended simply as sound introductions to neuroscience, neuropsychology, and philosophy of mind for those sufficiently motivated to go beyond my case stories and other meanderings.

Recent years have seen the publication of several fine works of popular science devoted to neuropsychology, neuroscience, and related topics. There are also some outstanding new textbooks. I can't think of a better general introduction to brain science than Ian Glynn's *An Anatomy of Thought: the Origin and Machinery of the Mind* (Weidenfeld & Nicolson, 1999). Susan Greenfield's *The Human Brain: A Guided Tour* (Phoenix, 1998) provides a useful brief survey; and for those who prefer to be spoon-fed, *Mind and Brain for Beginners* by Angus Gellatly and Oscar Zarate (Icon Books, 1998) is an entertaining and informative cartoon book. Bruno Aldaris's *Neuroscience for the Brainless* (Figment Books, 1994) which I quote in 'The Visible Man' is, sadly, virtually unobtainable these days.

Also recommended is *Phantoms in the Brain: Human Nature and the Architecture of the Mind.* (Fourth Estate, 1998) by V. S. Ramachandran and Sandra Blakeslee. Ramachandran is a remarkably inventive neurological thinker and, among many other topics, this stimulating book has some interesting things to say about phantom limbs and the neuro-biological bases of the body schema, complementing the impression-istic treatment I accord these topics in 'The Ghost Tree' chapters.

Todd E. Feinberg's *Altered Egos: How the Brain Creates the Self* (Oxford University Press, 2001) is another excellent collection of case-study vignettes and theoretical speculations. Among other fascinating material, it contains a chapter on the important (but in research terms relatively neglected) phenomenon of confabulation, which topic forms one of the strands of 'Soul in a Bucket'. I was amused to discover that we both appreciate the utility of the eye and pyramid symbol, though use it to quite different ends.

As for basic textbooks, I list my recommendations below. There may be other books of equal merit, but these are the ones with which I am familiar and they are very good indeed:

Gazzaniga, M. S., Ivry, R. B. and Mangun, G. R., *Cognitive Neuro-science: The Biology of the Mind*, Second Edition (Norton & Co., 2002).

Kolb, B. and Whishaw, I. Q., *An Introduction to Brain and Behavior* (Worth Publishers, 2001).

Rosenzweig, M. R., Breedlove, S. M. and Leiman, A. L., *Biological Psychology: An Introduction to Behavioral, Cognitive, and Clinical Neuroscience*, Third Edition (Sinauer Associates, Inc., 2002).

It is important to note that there are different varieties of neuro-psychology. One approach emphasizes *functional anatomy* and is con-cerned with studying the neural bases of psychological functions. Scientists working in this field are like mapmakers. Their mission is to explore the neurobiological landscape, charting the relationship between mental events and the structures and processes of the brain. Localization and distribution of functions is the central concern. This approach includes the classical method of examining the psychological consequences of 'focal' (localized) brain damage as well as the newer methods of cognitive neuroscience that use brain-scanning machines to explore patterns of activity in the normal, intact brain. The above-cited general texts are representative of this approach.

Other investigators – so-called cognitive neuropsychologists – have relatively little interest in the details of brain function. They assume, of course, that brain systems and mental life are intertwined, but their primary concern is the structure of the mind, not the brain.

Cognitive neuropsychologists study the performance of the damaged brain as a way of testing and refining theories of normal cognitive function. They start with a theory of cognition – some model of short-term memory, say, or of language production – and form hypotheses as to how memory or speech might be affected by brain disorder. The model gains support to the extent that patients' behaviour fits with predictions. Alternatively, their test performance may challenge the original model, leading to its modification or abandonment. What matters is the robustness (or otherwise) of the theoretical models rather than the precise nature of the underlying neurological disorder.

The best introduction to cognitive neuropsychology is *Human Cognitive Neuropsychology: A Textbook with Readings* (Psychology Press, 1996) by Andy Ellis and Andy Young. Alan Parkin's *Explorations in Cognitive Neuropsychology* (Blackwell, 1996) is also excellent, and *The Handbook of Cognitive Neuropsychology* (Psychology Press, 2000), edited by Brenda Rapp, is a useful source book.

The biological and psychological approaches are complementary. Although considerations of anatomy and physiology may be secondary to their main enterprise, the fruits of the cognitive neuropsychologists' labours are directly relevant to an understanding of the structure and functions of the brain. In order to know how psychological functions are represented in neural systems the functions themselves must be clearly delineated.

Modern clinical neuropsychology draws on both traditions, a fact nicely illustrated in David Andrewes's comprehensive survey, *Neuropsychology: From Theory to Practice* (Psychology Press, 2001). I also recommend *The Blackwell Dictionary of Neuropsychology* (Blackwell, 1996), edited by J. G. Beaumont, P. M. Kenealy and J. C. Rogers. Much

PAUL BROKS

more than a dictionary, this is an excellent reference work containing substantial entries on most aspects of clinical and experimental neuropsychology. The list of contributors is impressive.

For those specifically interested in psychiatric aspects of neurological disorder the classic text is W. A. Lishman's *Organic Psychiatry*, Third Edition (Blackwell Science, 1997). Although geared for a specialist readership, it is sufficiently lucid and digestible to be of interest to the motivated lay reader.

The emerging discipline of cognitive neuropsychiatry puts the principles and methods of cognitive neuropsychology to work in the field of psychiatric research. Peter Halligan and John Marshall's *Method in Madness: Case Studies in Cognitive Neuropsychiatry* (Psychology Press, 1996) provides a stimulating introduction. It includes a fine chapter on Cotard's syndrome by Andy Young and Kate Leafhead ('Betwixt Life and Death: Case Studies of the Cotard Delusion'), which helped shape my thinking as I came to write 'I Think Therefore I Am Dead'.

Themes of consciousness, self, and personal identity thread right through this book. I hope that professional philosophers will not find my excursions into their territory too naïve or superficial. If that is the case I blame the following: David Chalmers, *The Conscious Mind: In Search of a Fundamental Theory* (Oxford University Press, 1996); Francis Crick, *The Astonishing Hypothesis* (Simon and Schuster, 1994); Daniel Dennett, *Consciousness Explained* (Penguin Books, 1993); Owen Flanagan, *The Problem of the Soul: Two Visions of the Mind and How to Reconcile Them* (Basic Books, 2002); Gerald Edelman and Giulio Tonini, *Consciousness: How Matter Becomes Imagination* (Penguin, 2001); Nicholas Humphrey, *A History of the Mind* (Chatto and Windus, 1992); Colin McGinn, *The Mysterious Flame: Conscious Minds in a Material World* (Basic Books, 1999); Thomas Nagel, *The View from Nowhere* (Oxford University Press, 1986); John Searle, *The Mystery of Consciousness* (Granta Books, 1997); and Max Velmans, *Understanding Consciousness* (Routledge, 2000).

It would take another long chapter to summarize the range of ideas represented in these works (materialism, dualism, neural Darwinism,

Mysterianism, and so on), and I have resisted the temptation to offer such a summary – for one thing I don't feel qualified to undertake the task. Searle's highly readable book summarizes the thinking of some of the key figures in the current debate, as well as presenting the author's own views. If nothing else, it is worth reading for the testy exchange between Searle and Dennett in which each accuses the other of 'intellectual pathology'. It took me straight back to the school playground (*Fight! Fight! Fight!*).

I find Searle's 'biological naturalism' hard to fathom, but generally have difficulties fixing my own co-ordinates when it comes to the problem of consciousness. In 'Right this way, Smiles a Mermaid' the narrator stands accused of being a 'Mysterian'. Owen Flanagan, I believe, originally coined the term in honour (not quite the right word) of Colin McGinn for propounding the view that the mind-body problem is insoluble, or at least that we feeble-minded humans are incapable of solving it. I am much drawn to McGinn's deeply subversive position, but also find Dennett persuasive (as should be apparent throughout the book). That's my problem.

For an incisive analysis of the neuropsychological bases of consciousness, see Larry Weiskrantz's *Consciousness Lost and Found: A Neuropsychological Exploration* (Oxford University Press, 1997). The work of the neurologist Antonio Damasio has also been influential in shaping my thoughts on consciousness, self, and related matters. See, in particular, *Descartes' Error: Emotion, Reason and the Human Brain* (Picador, 1995), and *The Feeling of What Happens: Body, Emotion and the Making of Consciousness* (Heinemann, 1999). Joseph LeDoux's *Synaptic Self: How Our Brains Become Who We Are* (Macmillan, 2002) also makes an important contribution to our understanding of the neurobiological underpinnings of the self. Again, for those who prefer cartoon books, David Papineau and Howard Selina's *Introducing Consciousness* (Icon Books, 2000) is a sparky introduction to the field of consciousness studies.

In a recent collection of essays, *Consciousness and the Novel* (Secker and Warburg, 2002), the novelist and critic David Lodge offers some valuable insights concerning the representation of consciousness in literature. Works of literature – in contradistinction to science – describe 'the dense specificity of personal experience'. Science, from its objective, third-person perspective, tries to formulate universally applicable, general explanations. The subjective and the unique are anathema to science. Lodge suggests that 'Lyric poetry is arguably man's most successful effort to describe qualia' (the 'raw feels' of conscious awareness). 'The novel is arguably man's most successful effort to describe the experience of individual human beings moving through space and time.'

'To Be Two or Not to Be' draws heavily on the ideas of the philosopher, Derek Parfit, and I took the liberty of using the name 'Derek' for one of the central characters. Although most of what the fictional Derek has to say is, I believe, representative of the real Derek Parfit's ideas, there may be instances where the views of the two Dereks diverge in more or less subtle ways. The best way to become acquainted with Mr Parfit's thoughts on personal identity is to consult his masterwork, *Reasons and Persons* (Oxford University Press, 1984). Don't expect to read it in one sitting, however. A digestible account of some of Parfit's ideas – and much else besides – can be found in Jonathan Glover's *I: The Philosophy and Psychology of Personal Identity* (Penguin Books, 1991).

I saw Parfit lecture on personal identity, around the time that *Reasons and Persons* was published, in the Department of Physiology at Oxford during a series of seminars on the science and philosophy of mind. I thought at the time that he was saying something that was either quite trivial (if entertaining) or extremely profound and not a little disturbing. With the passage of the years I can see it was the latter. I have to say that, like his fictional counterpart, he did have a slightly wild look about him, and he was swigging water from a milk bottle throughout his presentation.

The scientific paper to which I refer in 'The Ghost Tree (1)' was co-authored with Andy Young and others and published in the journal *Neuropsychologia* (P. Broks, A. W. Young, et al., 'Face processing impairments after encephalitis: amygdala damage and the recognition of fear', *Neuropsychologia* 36, pp.59–70, 1998). In 'The Ghost Tree (2)' I refer to the concept 'the social brain' – the notion that the brain has evolved systems dedicated to social perception and understanding. A good general introduction to this way of thinking is *Friday's Footprint: How Society Shapes the Human Brain* (Oxford University Press, 1997) by Leslie Brothers. Simon Baron-Cohen makes a strong case for the relevance of the social brain to an understanding of autism in *Mindblindness* (MIT Press, 1995).

'The Sword of the Sun' was inspired by a story of the same name in Italo Calvino's *Mr Palomar* (Vintage, 1999), translated from the Italian by William Feaver . In 'Vodka and Saliva' I quote from Paul Ekman's *Telling Lies: Clues to Deceit in the Marketplace, Politics and Marriage* (Norton, 2001). 'The Dreams of Robert Louis Stevenson' was inspired by 'A Chapter on Dreams', which can be found as an appendix to *The Strange Case of Dr Jekyll and Mr Hyde* and *Weir of Hermiston* (Oxford World's Classics, 1998). The introduction by Emma Letley was very helpful, as was Robert Mighall's introduction to the Penguin Classics edition of *Jekyll and Hyde* (Penguin, 2002). 'The Visible Man', of course, pays homage to Kafka's classic allegorical tale 'Metamorphosis', available in *The Complete Short Stories of Franz Kafka* (Vintage Classics, 1999). 'In the Theatre' takes as its focal point Dannie Abse's disturbing poem, 'In the Theatre (A True Incident)', from his *Collected Poems, 1948-76* (Hutchinson, 1977). I had already drafted 'The Story of Einstein's Brain' when I came upon *Driving Mr Albert: A Trip Across America with Einstein's Brain* (Abacus, 2002), Michael Paterniti's entertaining account of his quest for the great man's grey matter and subsequent journey through America with it stashed in the trunk of his Buick. In 'Gulls' I refer to Milan Kundera's *Immortality* (Faber and Faber, 1991), translated from the Czech by Peter Kussi. In 'Soul in a Bucket' I mention Tom Wolfe's essay, 'Sorry, but Your Soul Just Died', which appears in *Hooking Up* (Jonathan Cape, 2000).

The Working Brain by Alexander Romanovich Luria (Penguin, 1973) is out of date, hard to obtain, and difficult to read – at least I found it so when I first came across it in my undergraduate days. I can't, for these reasons, recommend the book as an introductory text, but I cite it because its author has been a significant influence on my own approach to clinical practice. Luria, to whom I refer more than once in these pages, is one of the undisputed giants of neuropsychology, and *The Working Brain* (published four years before he died) summarizes his life's work. It presents a general theory of the organization of brain function – the distillation of forty years' work by Luria and his collaborators – as well as a comprehensive survey of existing knowledge of the classical domains of interest: the brain bases of perception, language, memory, thought, and action. Anyone developing a serious interest in the subject should at some stage track down, read, and appreciate Luria for the breadth of his vision of brain science, his recognition that the study of brain function is a multidisciplinary enterprise, and his insight that, ultimately, it becomes necessary to consider the brain in relation to other brains if one is to comprehend its workings: neuropsychology has social dimensions as well as biological and psychological. 'The eye of science,' he wrote, 'does not probe a "thing", an event isolated from other things or events. Its real object is to see and understand the way a thing or event relates to other things or events.'

Luria should be read not least for his understanding that neuropsychology concerns individual human beings – patients – struggling to make their way in a world rendered difficult and sometimes disturbingly strange by their brain damage. As far as neuropsychology is concerned, he was an advocate of 'romantic science' – recognizing the importance of combining close observation of individual patients (and understanding them as people) with a more systematic, 'classical' understanding of the facts of neurological disorder derived through conventional scientific method. This should be the aim of all clinicians – not to lose sight of the unique in the context of the universal, and vice versa.

Luria's scientific biography, *The Making of Mind: A Personal Account of Soviet Psychology* (Harvard University Press, 1986), edited by Michael and Shelia Cole, is a fascinating fusion of the personal, the political, and the scientific. But he is best known as a master of the extended case history, classic examples of which are *The Man with a Shattered World* (Penguin, 1975) and *The Mind of a Mnemonist* (Harvard University Press, 1986).

Another master of that genre is, of course, Oliver Sacks, whose works I have studiously avoided while writing this book. His influence was strong enough. I especially recommend *The Man Who Mistook his Wife for a Hat* (Picador, 1986) and *An Anthropologist on Mars* (Picador, 1995). These collections of neurological case histories, full of warmth and insight, embody the spirit of Alexander Luria, whom Sacks acknowledges as an important influence. Like Luria, he appreciates the complementarity of 'classical' and 'romantic' modes of understanding.

In the mid-1980s I was, for reasons not worth going into, somewhat disillusioned with clinical psychology. I thought about an academic career as a possible alternative, but did not relish the uncertain prospect of drifting on to the trail of short-term, post-doctoral research appointments with no guarantee of a proper job at the end. I might easily have given up neuropsychology altogether at that point. I might have been happy enough doing other things. Then I got a call from Merck Sharp and Dohme, the drug firm ('America's Most Admired Company' according to *Fortune Magazine* – I still have the commemorative mug). The unexpected call came from Susan Iversen, at that time Director of Behavioural Pharmacology at the MSD Neuroscience Research Centre. They were setting up a clinical research unit, she said, and would I be interested in joining them? I told her I knew nothing about psychopharmacology. 'Don't worry, love,' she said, 'you'll soon pick it up.' The advice was sound, and it is my advice to anyone interested in finding out more about neuropsychology, but wary of what might appear to be a rather daunting academic discipline: *Don't worry, love, you'll soon pick it up.*

ACKNOWLEDGEMENTS

Thanks to the patients whose stories lie at the heart of this book and from whom I have had the privilege of learning about the human dimension of neuropsychology, as opposed to what the textbooks teach. I hope I have given something in return.

It is a plain fact that this book would not have been written without the vision of my editor and publisher, Toby Mundy. The project was Toby's idea in the first place and he has seen it through with great élan. Apart from anything else, I thank him for his patience. It has also been a pleasure working with the impressive Bonnie Chiang, and, previously, Alice Hunt at Atlantic Books. I am grateful to Ian Pindar who read the penultimate draft in full and made numerous fine adjustments. Warm thanks also to my friends at *Prospect*, in particular David Goodhart and Alex Linklater, for the opportunity to write a monthly column for a very fine magazine.

From the bottom of my heart, I thank my entire family for their unfailing support down the years, but most of all my wife, Sonja, and sons Daniel and Jonathan, to whom I dedicate this book. You are more precious to me than ever.